Springer Series in Chemical Physics

Volume 124

The Springer Series in Chemical Physics consists of research monographs in basic and applied chemical physics and related analytical methods. The volumes of this series are written by leading researchers of their fields and communicate in a comprehensive way both the basics and cutting-edge new developments. This series aims to serve all research scientists, engineers and graduate students who seek up-to-date reference books.

More information about this series at http://www.springer.com/series/676

Vasily Artemov

The Electrodynamics of Water and Ice

Springer

Vasily Artemov
Center for Energy Science and Technology
Skolkovo Institute of Science and Technology
Moscow, Russia

ISSN 0172-6218 ISSN 2364-9003 (electronic)
Springer Series in Chemical Physics
ISBN 978-3-030-72423-8 ISBN 978-3-030-72424-5 (eBook)
https://doi.org/10.1007/978-3-030-72424-5

This Springer imprint is published by the registered company Springer Nature Switzerland AG
The registered company address is: Gewerbestrasse 11, 6330 Cham, Switzerland

Foreword

Water is a fascinating liquid and is the critical component of all forms of life on this planet. Both water and ice have anomalous structures, which have always been the subject of intense study. It is the ensuing unusual properties of water, which enable stability of many biological structures of both proteins and lipids to be maintained but also contribute in a large way to their function. Many physicists, biophysicists, biochemists, and chemists have been aware of this for a very long time and although the structure and properties of water have been discussed in numerous texts, there is a dearth of texts dealing solely with the structure and properties of water and ice. As an electrochemist, I am particularly aware that water is an essential component to many electrochemical reactions playing a large part also in biological charge transfer, so a text which brings together a synthesis of the nature of water and ice and their electrical properties as well as their interaction with electromagnetic radiation is particularly timely. This book by Vasily Artemov entitled "The Electrodynamics of Water and Ice" covers this whole subject in an innovative and stimulating way with elegant historical introductions to each chapter, which positively distinguishes this text.

The book is divided into five chapters. The Chap. 1 is a historical review of the structures of water and ice covering all the basic work associated with this including the radial distribution function, pH of solutions and conductivity, molecular-dynamic modeling of water and ice, with a very good summary at the end. Chapter 2 continues this theme dealing with the interaction of electromagnetic waves with water. Dielectric spectroscopy is included here and of course encompasses dielectric relaxation and the static dielectric constant of water connected with the water's electrical structure and properties. This was especially interesting from my point of view since I had done extensive work on the dielectric relaxation of lipid films in water. Dr. Artemov then follows this logically with Chap. 3 on the interaction of electromagnetic radiation with ice and then Chap. 4, which deals with the microscopic interpretation of dielectric properties of water and ice. The final Chap. 5 discusses the electrodynamics of electrolyte solutions and their conductivity. Particularly "eye catching" are the sections dealing with the electrodynamics of water and confined water and the behavior of water in strong electric fields.

I studied the behavior of phospholipids in strong electric fields for years so I found this chapter particularly informative.

Dr. Vasily Artemov has large experience of studying water on both theoretical and experimental levels. He was heading the lab of intermolecular interaction spectroscopy for many years before he joined the Skoltech Center for Energy Science and Technology in Moscow. He has had several stays in Germany, in particular, at the 1. Physikalisches Institut, Universität Stuttgart. His research interests and literature output correspond closely to the contents of this book. I became particularly interested in Dr. Artemov's work when he presented evidence in a seminar on the enhanced conduction of protons at solid/liquid interfaces, the mechanism of which he discusses in the last chapter of this book. This is extremely topical since the enhanced lateral conduction of protons at lipid/water interfaces was discussed and evidenced in relation to Peter Mitchell's Chemiosmotic theory. This exemplifies how relevant this book is to biological mechanisms of energy and ion transfer, which incidentally the Author highlights in relation to nerve function. As suggested by the Author, the ionic model of water paves the way for explaining many water-related anomalies.

Leeds, UK
February 2021

Andrew Nelson

Preface

Water is not only a substance No.1, but also an exceptional media of electromagnetic waves propagating on scales from the molecular to the global. Chemical reactions in aqueous solutions, nerve pulse transmission, and global communication systems are only a few of the many fields, which are significantly determined by the electromagnetism of water and ice. The dielectric properties of water and ice are equally important for many aspects of physical chemistry, soft-matter physics, electrochemistry, atmospheric sciences, radiophysics, medicine, biophysics, and neurosciences. The electrodynamic properties of water are accounted for in acid–base reactions, solvation processes, electric double layers, and in systems where water plays a role of solvent or reference liquid. The dielectric properties of ice, liquid water, and water vapor are also crucial for location and navigation, as they directly determine the quality of radio communications. In addition, such atmospheric phenomena as thunderstorms, lightning, precipitation charging, surface condensation, and evaporation depend on the unique dielectric properties of water and ice. Finally, people wonder how microwave ovens work, why water is transparent, how and why it conducts electricity, and how biological systems communicate. In spite of the fact that we would like to use the properties of water for many practical purposes, the electrodynamics of water and ice still do not have an exhaustive description, and is also poorly studied from a microscopic point of view.

In this book, I address a wide range of questions related to the electromagnetism of water and ice. The book contains a variety of experimental data on the dielectric properties of water in all its thermodynamic forms. The data are collated and ultra-broadband dielectric responses are combined in the unified terms for liquid water, ice, water vapor, and aqueous solutions. The interaction of electromagnetic waves with water is analyzed through the prism of microscopic molecular dynamics. The mechanisms behind different features of the water dielectric spectrum are discussed in fine detail. Well-established classic views on the structure of water and ice are critically analyzed and accompanied by new ideas. The physicochemical concepts of the electromagnetism of water and ice are formulated considering them on the same footing. Atomic-molecular dynamics are discussed in the context of microfluidics, atmospheric sciences, and electrochemistry.

In five interconnecting chapters, the reader will find a detailed explanation on why and how molecules of water are connected, and which molecular dynamics provide the variety of electrodynamic properties that water demonstrates in nature and technology. Although on the surface the book is dedicated to a narrow topic, the discussed questions go far beyond the electromagnetism of water and ice and provide a good platform for further investigations and interdisciplinary research of this vital, but still poorly understood compound.

Moscow, Russia
December 2020

Vasily Artemov

Chapter 1
A Historical Review of the Structures of Water and Ice

Abstract The details of the structure of water are important for understanding its dielectric properties and vice versa, because the nuclei and electrons of water particles are, in essence, charges interacting with each other and with external fields. Although water has been studied like no other substance, its atomic-molecular dynamics (both individual and collective) are still under discussion. Bernal–Fowler 100-year-old paradigm, based on early diffraction data, requires critical analysis. In this chapter, I consider the history of the notion of "water structure," which includes several milestones related to the appearance of particular experimental and theoretical methods. I discuss early electrical conductivity measurements, X-ray diffraction and neutron scattering techniques, pH measurements, isotopic tracer diffusion, nuclear magnetic resonance, and simulations of molecular dynamics. All these techniques and methods have significantly contributed to the development of the concept of hydrogen bonding in water and related substances. However, recent studies have shown that classic models still fail to reproduce the basic electrodynamic parameters of water and ice, and need to be improved.

1.1 What is "Structure" for Liquid Water?

The concept of "structure" for a liquid differs from that which is usually applied to solid crystalline materials. At first glance, a liquid is a fluid that has no shape without a container or confining medium, so it seems to have no "structure" at all, like a gas. Liquids are also essentially incompressible, maintaining the intermolecular distance and having a volume that is nearly as constant as in solids. One could say that this is due to the translational movement of its molecular species, a movement that is forbidden for ice molecules. However, glaciers, which are made of ice, slide and move, and this implies a translational degree of freedom, but this process appears on the much larger timescale. So, what does "structure" means for water and ice? Although water, strictly speaking, refers to the gaseous, liquid, or solid forms, *water* is used in the text to refer to *liquid water*, unless otherwise stated.

Although long-range order is not observed in water and is only a feature of carefully prepared ice samples, the concept of short-range order, in contrast, is fully

V. Artemov, *The Electrodynamics of Water and Ice*, Springer Series in Chemical Physics 124, https://doi.org/10.1007/978-3-030-72424-5_1

applicable to both forms. The short-order structure is mainly formed by pair interactions of water molecules, which are identical in ice and water. The local molecular environment is also expected to be nearly the same. Pair interactions, and the corresponding local water structure, can be tested by simulations of molecular dynamics[1] and by the adjustment of the potential of the interaction of water molecule pairs. Abascal et al. [1] introduced the most popular rigid four-point model of the water molecule (TIP4P/2005 model), which quantitatively reproduces the phase diagram and the density of water and ice. Using this model, Ramirez et al. [2] showed that the temperature (the thermal energy) of a system of water molecules defines the condition of the long-range order formation. Figure 1.1 shows two quasi-equilibrium configurations of TIP4P/2005 water molecules that represent the quasi-equilibrium "structure" of water and ice. These two configurations are obtained at two different thermal energies and both have nearly the same local molecular environment, while the "structure" looks different. This example shows that, although the pair potential stays the same, the residence time of molecule near the equilibrium position defines the condition of long-order formation.

Figure 1.1 shows quasi-equilibrium configurations. In real ice, however, each molecule has a chance to overcome the potential barrier and change the equilibrium configuration, which in the frame of ergodicity,[2] means that, when the observation period is long enough, the structure shown in Fig. 1.1b is equivalent to the structure shown in Fig. 1.1a (and vice versa). The difference between what we call the "structures" of ice and water is in the corresponding observation times and the period of the averaging of relative molecular configurations.

As the structure of water and ice depends on the time and spatial scales over which it is determined, we should define these characteristic timescales. Eisenberg and Kauzmann [3] suggested several types of water structures, depending on the time of averaging [3] and the experimental techniques used for their determination. In particular, they introduced the instantaneous or I-structure, which is used for the timescale less than 1 ps; the vibrationally averaged, or V-structure, for 1–10 ps; the diffusion averaged, or D-structure, for the timescale more than 10 ps; and the potential-energy-minimum static F-structure ($t \rightarrow \infty$). According to this classification, Fig. 1.1 represents the F-structure, as they were obtained by potential energy minimization and do not reflect the real instantaneous I-structure. The latter is an idealization, because no method can accurately measure it. For example, conventional electrochemical methods deal with D-structure, while infrared spectroscopy measures V-structure. Even neutron and X-ray scattering techniques, which have a shortest interaction time of a single photon/neutron with a sample, still have a

[1]Experimental methods of the short-order test are considered in Sects. 1.2 and 1.5.

[2]Ergodicity expresses the idea that any particle of a moving dynamic system will eventually visit all parts of the space that the system moves in. This implies that the average behavior of the system can be deduced from the trajectory of a selected particle. Equivalently, a sufficiently large collection of random samples from a process can represent the average statistical properties of the entire process.

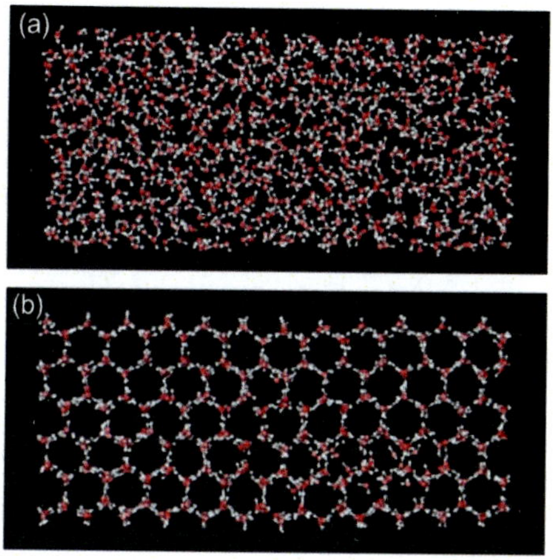

Fig. 1.1 Pictures of two quasi-equilibrium configurations of $N = 870$ water molecules obtained by minimizing the combination of Lennard-Jones and Coulomb interactions of pairs of TIP4P/2005-model water molecules. **a** The structure of water. The thermal energy is higher than that allowed for long-range order formation. **b** The structure of ice. The thermal energy is low enough for long-range ordering. The local molecular environment stays nearly the same in both cases. Adapted from [2] with permission from the PCCP Owner Societies

relatively large data acquisition time and cannot provide a true I-structure test. The challenge of recent decades is to theorize an I-structure of water, the dynamics of which would consistently give the V-, D-, and F-structures.

Some studies [4] show that there may be more than one meta-stable structure in water and ice, which coexist in dynamic equilibrium. This concept has been found useful for the explanation of neutron diffraction experiments [5], and gives a reason to talk about the concept of low-density and high-density liquid states. Nevertheless, a microscopic description of these two states is still lacking. It has recently been shown that the combination of low-temperature ordered states of water based on the tetrahedral structure and a high-temperature disordered state based on a distorted tetrahedron structure can qualitatively explain some anomalies of water [6]. However, the application of this approach for the explanation of the electrodynamic properties of water is limited, mainly due to the lack of understanding on how and why these two different water structures are formed, and change into each other.

1.2 Bragg Scattering and Bernal–Fowler Water

The first attempts to describe the dielectric properties of water on the microscopic level were undertaken by Röntgen [7] and Kohlrausch [8]. However, only the ideas of Bernal and Fowler [9] were elaborated enough to form a basis for the majority of the modern structural models of both ice and water. Below, we discuss the main arguments that have been used to develop this famous model of water structure.

1.2.1 Scattering of X-Rays by Liquid Water

Bernal and Fowler used X-rays to identify the local molecular environment of H_2O molecules in a condensed state. They assumed that water molecules have spherical symmetry with a triangular arrangement of the hydrogen and oxygen nuclei. Five characteristic points were chosen for this model of the water molecule based on earlier spectroscopic data [10]. The oxygen was placed at the center of the sphere as shown in Fig. 1.2. Two hydrogen atoms (protons) were placed in the y-z-plane about 1 Å from the oxygen atom, and 1.5 Å from each other so that the H–O–H angle was about 105°. Two more points, which represent electron densities that compensate for the density of the protons, were settled in the x-z-plane. This tetrahedral model molecule of H_2O became the structural unit of Bernal–Fowler water. It has been suggested that such a molecule has some preferential orientations with respect to surrounding molecules, due to electrostatic interactions between the specific points of its positively and negatively charged parts. The search for these preferred configurations became the main subject of Bernal–Fowler work and anticipated future studies of molecular dynamics.

Bernal and Fowler noticed that an experimental X-ray pattern (see Fig. 1.3) that was scattered by ambient water (cure 1 in Fig. 1.3b + open symbols) and the model spectrum of amorphous quartz with a tridymite structure (curve 3) have much in common.[3] In particular, the positions of the minima and maxima of the intensity distribution are approximately the same. Authors assumed that water, by analogy with quartz, has a coordination of atoms as in amorphous SiO_2, and suggested a quartz-like structure of water, shown in Fig. 1.4. The asymmetric non-linear shape of water molecules allowed them to form a tetrahedral molecular coordination as shown in Fig. 1.4a. The assembly of pentawater complexes gives an open hexagonal structure shown in Fig. 1.4b, which is similar to the structure of quartz shown in Fig. 1.4c.

In order to obtain the correct density of water, Bernal and Fowler assumed that the volume of voids between molecules is approximately equal to the occupied volume. As a result, a good coincidence between the experimental curve 1 and the theoretical curve 3 was achieved. In such a way, the tetrahedrally coordinated arrangement of water (see Fig. 1.4a), self-assembled into hexagonal rings, was introduced to the water model as the most suitable configuration of the hydrogen and oxygen atoms satisfying the X-ray data. However, this model structure corresponds to the most probable position of atoms (F-structure) and does not account for any molecular dynamics.

Although a good coincidence of Bernal–Fowler model with X-rays diffraction data has been achieved, the quartz-like water model is static by nature. Electro-diffusion data, rheological properties, and thermal effects were missing. Assuming that water consists of long-lived H_2O molecules similar to those observed in steam with small mutual deformations, the model offers no explanation of the experimentally observed

[3] Note that in this kind of experiment, water atoms are not directly observed, because X-rays interact with the electronic density around the oxygen atoms, but not the atoms themselves.

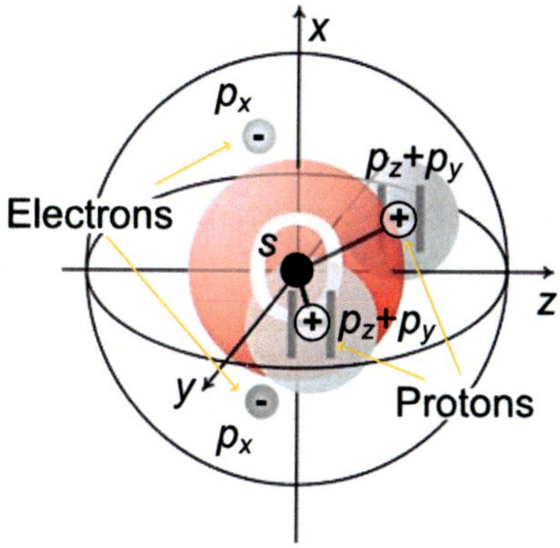

Fig. 1.2 The electron and nuclei distribution in a water molecule according to the Bernal–Fowler model. Adapted from [9], with the permission of AIP Publishing

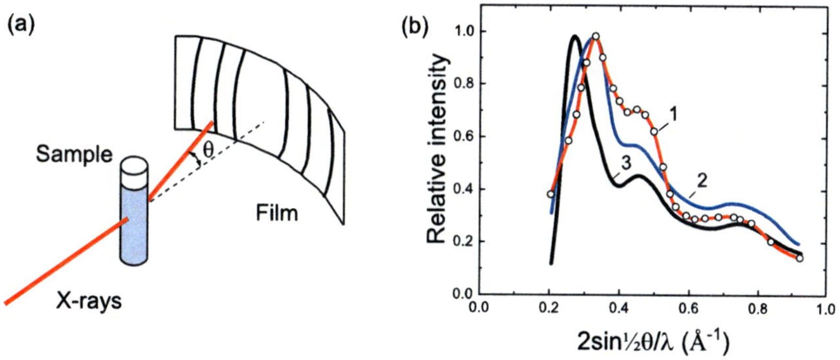

Fig. 1.3 **a** The schematic of X-ray experiments with water. **b** X-ray diffraction intensity-angle distributions for water (curve 1); theoretical pattern for a quartz-like water model (curve 2); and the theoretical pattern for amorphous quartz (curve 3). Adapted from [9], with the permission of AIP Publishing

high mobility of H^+ and OH^- ions,[4] which assumes something other than the simple hydration of ionic species, a mechanism that allows H^+ and OH^- ions to move through the media and impart coherence.

Therefore, a number of modifications to the Bernal–Fowler water model were later proposed. For example, Fig. 1.5 assembles the main species that different models introduced to the quartz-like water model in order to interpret the scope of electro-diffusion data. Defects of different types can be found: vacancies, interstitials, chains, excess protons, holes, free, and randomly bounded molecules. All these particles were needed to provide additional degrees of freedom to the static lattice structure revealed

[4]The electrical mobility of H_3O^+ and OH^- ions is known to be almost an order of magnitude larger than the mobility of other ions with the same charge such as Li^+, Na^+, and K^+ [11]. This point was later understood by the Grotthuss mechanism (see Sect. 1.3).

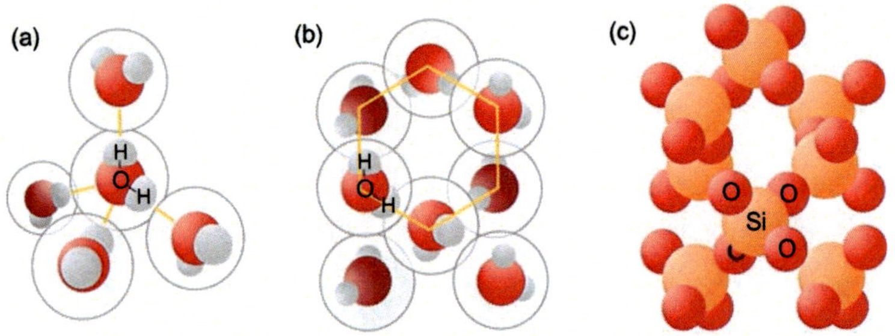

Fig. 1.4 Molecular configurations revealed from X-rays scattering data shown in Fig. 1.3b. **a** The tetrahedral coordination of water molecules. Four molecules form a tetrahedron around a central molecule according to the charge density distribution shown in Fig. 1.2. **b** The "quartz" water structure suggested by Bernal and Fowler. **c** The structure of quartz that was used as a prototype

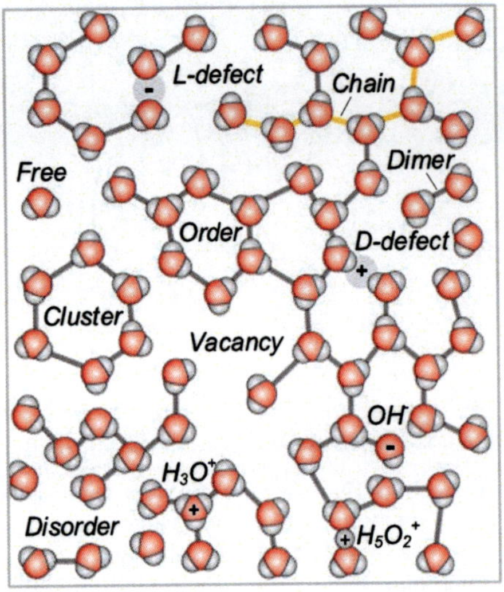

Fig. 1.5 The structure of water based on the Bernal–Fowler model with defects in the hexagonal structure that have been used by different authors to explain its dynamic properties. Compare with the structure according to the ionic model in Fig. 4.5

by X-ray diffraction. Interestingly, for the interpretation of some experimental data, such as electrical conductivity, dielectric relaxation, and the dielectric constant, all these defects become even more important than the molecular structure itself.

Several ideas, accounting for the heterogeneity of the structure of water, have been suggested to improve the Bernal–Fowler model. Concepts of light and heavy phases, ice-like clusters, a high concentration of short-lived ionic species, liquid–liquid transitions, polyamorphism, and the concept of high- and low-density water have been introduced [12–16]. However, the debates around these proposals continue, and no model yet has a decisive advantage. Nevertheless, it is obvious that water is more than just an ideal tetrahedral arrangement of spherical molecules shown in Fig. 1.4. There is much more "disorder" in water and the exact nature, size, and sharpness of the boundaries between defects and fluctuating regions are still needed to be clarified [14]. The experimental and theoretical efforts of recent years using intense X-rays from synchrotrons (SAXS) and from free-electron lasers (XFEL) look promising [17]. These techniques have much shorter exposure time than in the original Bragg experiments, and thus they are getting closer to the I-structure.

A century of X-ray studies after Bernal and Fowler has shown that the short-order arrangement in water is similar to that in ice crystals. However, the microscopic dynamics, which cannot be studied by this method, were still missing. In order to introduce dynamics to Bernal and Fowler's concept, Frenkel (1946) proposed the idea of the simultaneous vibrational and translational motion of molecules, which is suitable for liquids [18]. The idea was that the oscillations of the molecules in quasi-crystalline cells have been assumed to be similar to those in solids, but the translational displacements of the cell as a whole with the corresponding free path were equal to the first step of diffusion (see Sect. 1.3 for details). Such dynamics were proposed to account for a distinctive feature of liquids, and were well suited to the molecular dynamics in water. This model has crystalline and gas-like properties at the same time. This can explain the X-ray diffraction data and leaves space for dynamics. Frenkel's classical model was effectively used to interpret quasi-elastic neutron scattering data (see Sect. 1.2.3), but was not further developed because in the 1970s, the main focus was turned to simulations of quantum-mechanical molecular dynamics (see Sect. 1.6), where the calculated volume of water is too small to model Frenkel's dynamics, which require significantly higher computational cost than computers could afford.

1.2.2 *X-Ray Crystallography of Ice*

X-ray interaction with ice was historically studied separately from that for liquid water. The method dates back to the first systematic detection of the diffraction patterns of ice crystals by Rinne [19], and their interpretation in relation to the geometrical arrangement of oxygen atoms by Bragg [20]. A large inaccuracy remained in the arrangement of atoms and molecules in ice, until it was clarified by neutron scattering (see Sect. 1.5), and infrared and dielectric spectroscopy (see Chap. 2). Nowadays, although the average position of oxygen atoms has been defined in detail, there are still many questions concerning the details of ice's structure and atomic dynamics. The problem is that it is difficult to prepare a homogeneous ice sample, and then establish its crystal symmetry, which corresponds to the ice lattice. The scattered intensity of X-rays depends on the orientation of the sample relative to the incident beam, as shown in Fig. 1.6. Thus, a lot of initial assumptions are needed in the interpretation of X-ray diffraction data. The construction of the model is additionally complicated by the fact that the crystal growing process is time dependent, and the crystal structure is significantly dependent on the boundary conditions.[5] Never-

---

[5] It should be noted that a single crystal of ice Ih (hexagonal) cannot appear upon cooling a hypothetical infinite volume of water, since the latter does not have a distinguished axis. Ideal ice Ih crystals grow on cooled surfaces with a surface-dependent structure (see, for example, [24]).

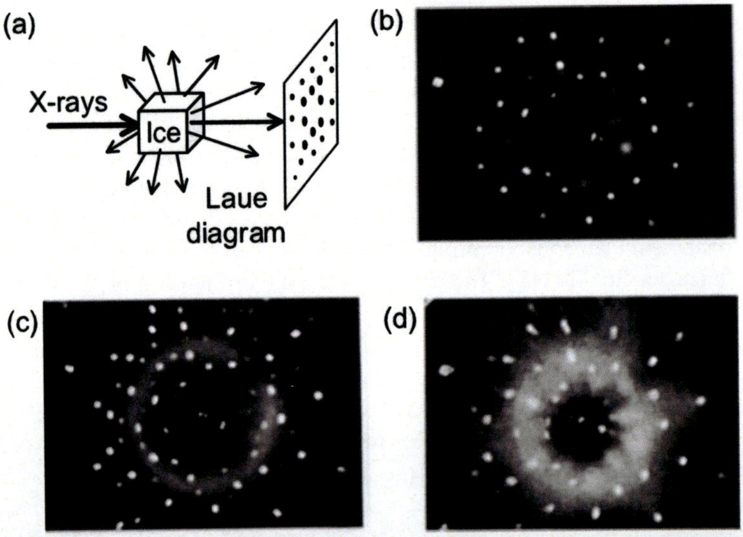

Fig. 1.6 X-ray diffraction experiments with ice crystals. **a** The experimental schematic. **b**, **c**, and **d** Laue diagrams of ice crystals for an X-ray beam directed at an angle of 4.6 °C to the *c*-axis of the crystal at −5 °C, at −196 °C (liquid nitrogen), and a temperature rising from −180 to −5 °C, respectively. Parts (**b**) to (**c**) adapted from [23] with permission of the International Union of Crystallography

theless, X-ray crystallography of ice provided useful information about the relative position of water molecules, averaged over long periods of time (usually more than 1 s [21]).

Unlike liquid water, ice crystals scatter X-rays non-uniformly, producing areas of low and high intensity as shown in Fig. 1.6a. Scattering patterns from ice crystals are known as Laue diagrams,[6] shown in Fig. 1.6b–d. Similar to water, the interpretation of these patterns requires knowledge about the structure, which for ice reduces to the knowledge of the crystal lattice structure, which is unknown a priori and cannot be deduced directly from the geometrical shape of the sample. Everyone knows that the shapes of snowflakes vary significantly depending on the external conditions, not to mention the more than 19 forms (both stable and meta-stable ones) of bulk ice that one can find on the phase diagram of water (see [22]).

In order to describe the structure of ice, Bernal and Fowler relied on the infrared and Raman spectra, which show common vibrational spectral features for ice, liquid water, and water vapor (see Fig. 2.13). As the latter was supposed to consist of H_2O molecules, the condensed forms of water were assumed to be an assembly of H_2O

[6]Named after Max von Laue, who used diffraction from crystals to prove the short wavelength of X-rays.

molecules too.[7] The 3D structure of ice was deduced from general assumptions that molecules with the structure shown in Fig. 1.2 electrostatically interact with each other. As the interaction of X-rays with ice reduces to the interaction with oxygen atoms (except modern small-angle X-ray diffraction studies, which are sensitive to hydrogen atoms as well [25]), no hydrogen–oxygen configurations leaving the positions of oxygen atoms unchanged have been rejected on the ground of X-ray experiments.

Figure 1.7 shows the structure that Bernal and Fowler proposed for ice on the basis of the least energy configuration of five-point tetrahedral H_2O species. This structure has been used for the interpretation of Laue diagrams of ice. The oxygen atoms in this model form a hexagonal configuration, similar to that for liquid water discussed above. One oxygen atom was assumed occupying each corner of the hexagon. The hexagons, being assembled in a plane, form a wrinkled surface of conjugate rings. Then planes alternate in an (a)–(b)–(a)–(b) order along the c-axis, where each (b)-plane is a reflection of the (a)-plane along the same axes as the planes themselves [26]. The distance between two neighboring (a)- or (b)-planes is 7.3 Å. The distance between any neighboring oxygen atoms is about 2.8 Å. The whole structure has hexagonal symmetry ($P6_3/mmc$ structure group[8]). The angle between the sides of the hexagons is close to the tetrahedral angle (109.5°) and the angle between the hydrogen atoms in the water vapor molecule (104.5°). The cavities inside the hexagonal rings were assumed large enough to place another water molecule inside. That is why the electrical conductivity and self-diffusion coefficient of ice along the c-axis, and perpendicular to this axis differ several times, as has been experimentally confirmed latter [27, 28].

The hydrogen atoms are not shown in Fig. 1.7, but they were placed by Bernal and Fowler in fixed positions on the lines between oxygen atoms. Later Pauling adopted this configuration, because he found that it does not contradict the residual entropy of ice at low temperatures [29]. On the basis of statistical physics, Pauling then deduced the following phenomenological ice rules, which although not applicable for real ice crystals, were adopted, since they directly followed from the structure of the water molecule proposed by Bernal and Fowler (see Fig. 1.2):

- Each oxygen atom has two hydrogen atoms attached to it at distances of about 0.95 Å, forming a water molecule, the HOH angle being about 105° as in the gas molecule.
- Each water molecule is oriented so that its two hydrogen atoms are directed approximately toward two of the four oxygen atoms which surround it tetrahedrally.

[7]Modern data shows that the infrared and Raman spectra of ice and water are more complex than those for water vapor. Apart from line shift and broadening [31], overtones, Fermi resonances, and contributions from ionic species have been observed [32] (see Chap. 2 for details).

[8]$P6_3/mmc$ means having a sixfold screw axis (rotation around an axis in addition to a translation along the axis); the structure repeats itself three times during a complete 360° rotation; the "m's" stand for mirror planes perpendicular to the basal plane and parallel to the "c"-axis, and with the c in "mmc" standing for the glide plane.

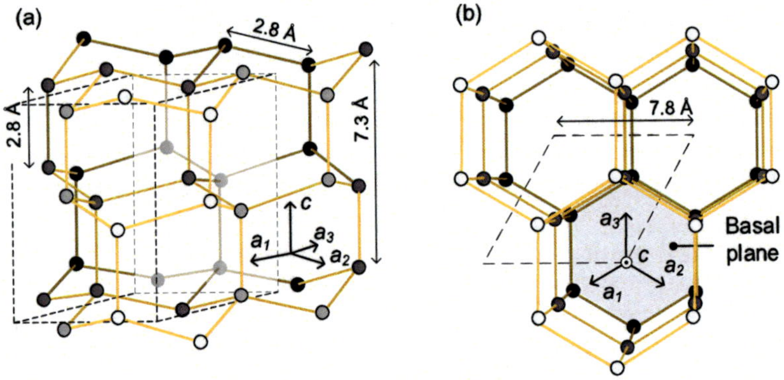

Fig. 1.7 Structure of ice Ih according to X-ray diffraction data. **a** Structure viewed perpendicular to the crystallographic c-axis. **b** Structure viewed along the c-axis. Dots show the centers of oxygen atoms. Atoms that are closer to the viewer have a lighter tone. The dashed lines outline the unit cell. The straight lines between the atoms are guidelines for the eyes

- The orientations of adjacent water molecules are such that only one hydrogen atom lies approximately along each oxygen–oxygen axis.
- Under ordinary conditions the interaction of non-adjacent molecules does not appreciably stabilize any one of the many configurations satisfying the preceding conditions with reference to the others.

At non-zero temperatures, ice crystals are free to change atomic configurations, for example, by synchronous proton exchange, as shown in Fig. 1.8. For this, Pauling [30] suggested two possible mechanisms of hydrogen atom migration: the rotation of molecules around the center of mass and the intermolecular transfer of the hydrogen nuclei.[9] Change from one of the two cyclic arrangements of hydrogen nuclei to the other was permitted by the postulates mentioned above. The fact that the dielectric constant of ice is higher than that of water shows that there are significant dynamics in the ice lattice, which give it the ability to polarize. The crystal changes the atomic relative arrangement under the influence of an electric field or due to thermal fluctuations. Pauling's concept of the residual entropy of ice means that at very low temperatures, close to absolute zero, ice does not become (in a reasonable period of time) a perfect crystal with no randomness of molecular orientation. Thus, the defects of an ideal ice structure can be responsible for the electric and transfer properties of ice even at very low temperatures.

Although recent studies continue to tweak the details of the quartz-like water model, the main features, and the positions of oxygen atoms, remain the same. However, the degree of tetrahedrality has been criticized [33, 34], and it is clear

[9]Note that latter is circled and seems equivalent to molecular rotation. This can be true from the static dielectric constant point of view. However, from the viewpoint of dynamic conductivity, the molecular reorientations are not equivalent. The migrating proton (p^+) produces local current, which contributes to the spectrum at high frequencies and gives, for example, microwave absorption (see Chap. 2 for details), while the molecular reorientation is not.

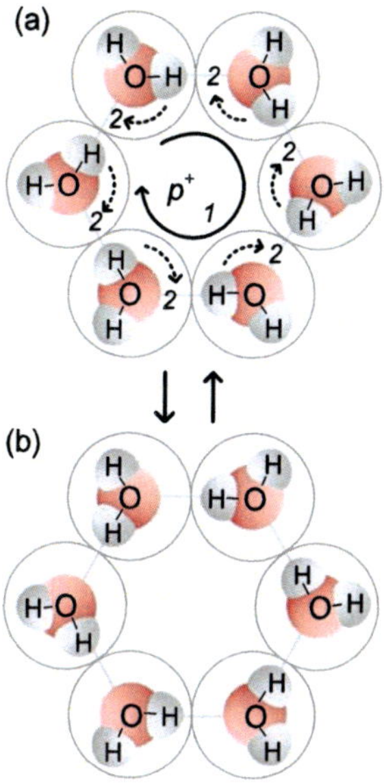

Fig. 1.8 Two equivalent molecular configurations of ice, from the X-ray diffraction point of view, with different configurations of hydrogen atoms (**a**) and (**b**). There are two ways to switch between these configurations. The first is shown by the large arrow (number 1), which demonstrates the direction of the proton transfer. The second is shown by small dashed arrows (numbers 2), which express the directions of molecular reorientations. Only one hydrogen atom is shown for each water molecule for clarity. Both mechanisms are equivalent from the viewpoint of initial and final arrangements of the atoms, but only the first scenario provides an intermolecular electric current

that real water is more disordered than the ideal tetrahedrally arranged openwork network shown in Fig. 1.7. A further model that can account for the interstitial-diffusion mechanisms and explain the similarity of the dielectric properties of water and ice (described in Chap. 4) is needed.

1.2.3 The Radial Distribution Function

A direct way to describe and compare the local molecular environments of water and ice on a sub-nm scale is the analysis of their radial distribution functions (RDFs). RDF defines the probability g_{ij} of finding an atom i at distance r from another tagged atom j (see Fig. 1.9a). In other words,

$$g_{ij}(r) = \frac{dn_r}{4\pi r^2 dr \cdot \rho}, \tag{1.1}$$

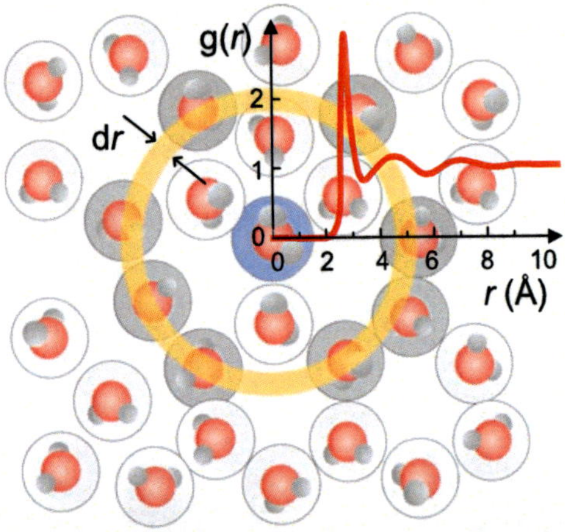

Fig. 1.9 The local environment of the reference water molecule (blue). Gray molecules are those whose centers are within the circular segment designated by dotted lines. The algorithm of the RDF (red curve) determination involves accounting for the molecules within a distance of r and $r+dr$ from a central particle

where dn_r is a function that computes the number of particles within a shell of thickness dr, as shown in Fig. 1.9. For water and ice, one can distinguish oxygen–oxygen, g_{OO}(r); oxygen–hydrogen, g_{OH}(r); and hydrogen–hydrogen, g_{HH}(r) pair distribution functions. However, one needs preliminary knowledge (a model) of the water structure, which can be used for the parallel analysis of X-ray and neutron scattering data, and, reconstructing the data, one faces a mathematical ill-posed problem [35] with multiple possible solutions. In addition, the methods of X-ray and neutron scattering provide an atomic-scale resolution, but involve significant space and time averaging over the sample size and the acquisition time, respectively. Thus, RDF is a model-dependent function, which reflects a diffusion-averaged structure.

As RDF cannot be measured directly and requires a preliminary assumption about the structure, a conventional RDF analysis assumes that water molecules are stable structural units, which do not disintegrate with time and form the preferential configurations prescribed by the structure of the individual molecule. Thus, the analysis follows the Bernal–Fowler general idea that water consists of H_2O, and is based on the search for different molecular configurations. At the current stage of research, data are collected with good statistical accuracy and generally converge [36]. However, the result of data processing still depends on the model, and the initial assumptions that were made. That is why there is still large uncertainty among the RDFs obtained by different groups [37]. To understand the problem, let us look at how RDFs are determined (Fig. 1.10).

The integral scattering intensity is assumed to have the following form:

$$I(Q) = \left\langle \left| \sum_{k=1}^{N} b_k \exp(i Q \cdot r_k) \right|^2 \right\rangle, \tag{1.2}$$

where b, r, and Q are the atom scattering length, the radius vector, and the scattering momentum, respectively, and the angular brackets denote ensemble averaging.

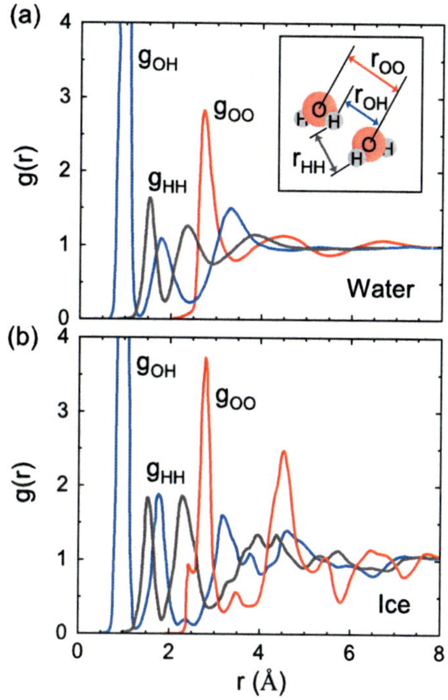

Fig. 1.10 Pair radial distribution functions (RDFs) for **a** water, and **b** ice: oxygen–oxygen, g_{OO}, oxygen–hydrogen, g_{OH}, and hydrogen–hydrogen, g_{HH}). The inset shows distances between different atoms of two neighboring molecules, which corresponds to the first maxima of the RDFs

With these assumptions, the scattering intensity per atom is represented by the additive sum:

$$I(Q) = I_{self}(Q) + I_{\text{intra}}(Q) + I_{\text{inter}}(Q), \tag{1.3}$$

where first two terms correspond to the molecular form factor $< F(Q)^2 >$, and the last term represents intermolecular correlations. It is important to note that if water molecules are short-lived, the total scattering intensity is a more complex function than that presented by (1.3), and does not converge to the additive sum. The term $<F(Q)^2>$ is assumed to be equivalent to that in water vapor, which can be obtained by X-ray diffraction of steam [38] in the Debye approximation [39].[10] The latter considers water as a system of molecules similar to those in the gas phase, but perturbed by mutual interactions.[11] In other words, one can write

$$\langle F(Q)^2 \rangle = \sum_{ij} x_i x_j f_i(Q) f_j(Q) \frac{\sin Qr_{ij}}{Qr_{ij}}, \tag{1.4}$$

[10]The Debye scattering equation allows one to calculate the scattered intensity from an isotropic sample which does not presume the periodicity of the underlying structure.

[11]It is shown in Chap. 3 that the intermolecular proton exchange which plays an important role in the electrodynamics of water and ice is missing in this approach.

where x_i is the atomic fraction of atoms of type i, r_{ij} are the intramolecular distances between the centers of nuclei.

At this level of approximation, the geometric structure of the water molecule is assumed to be a rigid sphere, and individual atomic scattering factors are calculated for isolated unbound particles. In fact, covalent bonding and the mutual interaction of molecules change the charge distribution and, therefore, the atomic form factors defined by (1.4).

The structural factor $S(Q)$ is determined from the molecular form factor by

$$S(Q) = \frac{I(Q)}{\langle F(Q)^2 \rangle}. \quad (1.5)$$

The contribution of intermolecular correlations to the total intensity (see the last term in (1.3)) is given by [33]

$$I_{\text{inter}}(Q) = \sum_{i \le j} (2 - \delta_{ij}) x_i x_j f_i(Q) f_j(Q) S_{ij}(Q), \quad (1.6)$$

where the correlation factor $S_{ij}(Q)$ between atoms of type i and j is directly related to the RDF by

$$S_{ij}(Q) = 4\pi\rho \int_0^\infty r^2 (g_{ij}(r) - 1) \frac{\sin Qr}{Qr} dr, \quad (1.7)$$

where ρ is the atomic density.

Therefore, the structure factor $S(Q)$ and the RDF itself can be calculated directly from (1.7). However, a preliminary structural model is needed to weigh the scattering correlations between the oxygen and hydrogen atoms. For this purpose, the diffraction pattern is analyzed by the empirical potential structure refinement [40]. Using molecular dynamics and ab initio methods, interatomic potential energy functions are consistently tuned to reconcile the simulated and measured structure factors $S(Q)$. The aim of the simulation is to find realistic constraints on the forces between atoms, inside and between molecules, which will fit the experimentally observed RDF. Since this problem generally has an unlimited number of solutions, the result always depends on the model used. This is why there are still some differences between the RDFs obtained by different authors depending on the initial assumptions and the potentials that were applied.

The first RDF for water was reported almost a century ago [41]. Figure 1.9 shows the modern RDFs for water and ice. The $g_{OO}(r)$ function for water shows well-defined maxima at 2.9, 4.5, 6.7 Å and minima at 3.4, 5.5 Å. The positions of the peaks have a ratio close to the 1.63, which is expected for a tetrahedral water structure [42]. However, they are broad and overlapping, showing a wider range of O–O–O angles than those expected for the ideal tetrahedral arrangement introduced by Bernal and Fowler. The exact degree of water tetrahedrality is still debated and the structural model of water, which corresponds to the X-ray and neutron scattering data, is still developing [33, 34, 36, 43]. Obviously, water is more disordered than the ideal tetrahedral arrangements of amorphous SiO_2. Though the RDF analysis roughly confirms the trigonal geometry of water molecules in the liquid state and shows that time-averaged four-coordinated structures dominante [12, 44].

Ice generally reproduces the data for water, but there are several additionally resolved maxima near 2.2, 3.4, 5.2 Å. These maxima show that ice has a long-order and a higher degree of tetrahedrality compared to liquid water. Why is the local diffusion-averaged structure of water different from that of ice?

An analysis of the fine structure of the water RDF made by Samoilov [45] showed that there is an additional smaller maximum at 3.5 Å, which is not expected from the purely tetrahedral arrangement of H_2O molecules. He assumed that this maximum corresponds to the interstitial molecules trapped in the hexagonal rings (see Fig. 1.9). Samoilov used a pair distribution function, defined as $4\pi r^2 \rho(r)$, where $\rho(r)$ is the density, which can be understood as the deviation of the local density of water in comparison with the averaged density. Figure 1.11 shows (a) the experimental and (b) the calculated distribution functions. A relative subtraction shows that there is an additional density peak near 3.5 Å, which does not appear in the ideal structure. Samoilov found that several percent of water molecules are temporarily trapped in the interstitial cavities. This assumption explains the deviation from the ideal structure of Bernal and Fowler, but does not explain why the interstitial states differ from the lattice states. Most recent studies of the radial distribution function [42] show that phenomenological constants allow a description of the experimental data with satisfying accuracy, and no further improvement of the experimental data is needed. However, there is still no simple structural model, which describes the RDF and satisfies the scope of the thermodynamics, electrodynamics, and structural properties of water.

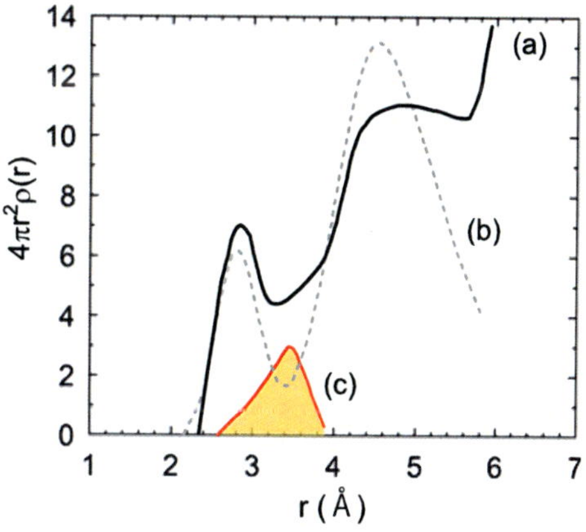

Fig. 1.11 A comparison of **a** experimental and **b** theoretical curves of the radial distribution of oxygen atomic density for water at 275 K. The function difference **c** shows a hidden density maximum at 3.5 Å, which is not expected from the ideal tetrahedral structure. It corresponds to the molecules trapped in the cavities of the hexagonal rings (interstitial molecules)

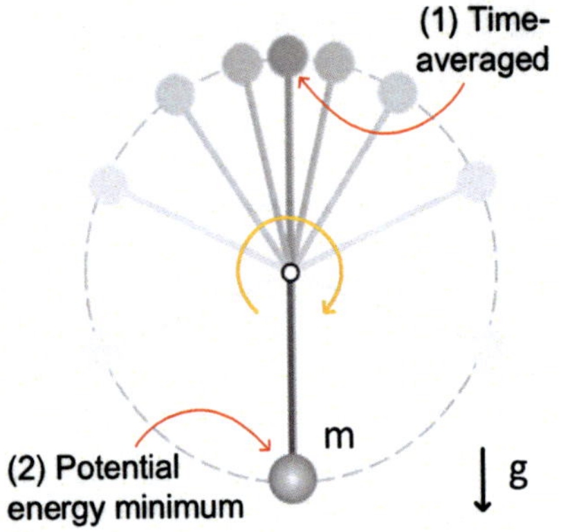

Fig. 1.12 The rotating-pendulum model that describes the difference between (1) the time-averaged position and (2) the potential-energy-minimum point. An inverted position is more likely at averaging as the pendulum spends there more time. However, it would never be observed if one searches the equilibrium by the minimization of potential energy (see the text for analogy with the molecular motion)

Three main problems prevent the understanding of the local structure of water by RDF analysis. First, the result is significantly determined by the initial assumptions, which are not yet well defined. As a result, we get significantly different results for RDF, depending on the molecular potential that we use. Second, the distribution function reflects the time- and space-averaged configurations of short-lived local molecular environments. This fact makes the analysis of the potential-energy-minimum F-structure controversial. For example, the averaged position for rotating pendulum (see Fig. 1.12) corresponds to the upper point, because it spends more time there than at any other point, while the potential-energy-minimum position is obviously at the bottom. This analogy shows the difference between the diffusion-averaged and potential-energy-minimum structures. Third, although the concept of interstitial molecules qualitatively explains the denser configuration of water in comparison with ice, it was not confirmed by high-pressure experiments [46]. The expected increase of the concentration of interstitial molecules at higher pressure in this model has not been observed experimentally.

Therefore, we cannot learn more about water structure using the RDF only. The RDF confirms the roughly tetrahedral arrangement of molecules, but does not provide the angular distribution among neighboring molecules, nor does it reflect the distribution around the local atom within short periods of time. Thus, the RDF analysis highlights the D-structure but says nothing about the I- or V- structures [3].

1.3 Direct Current Conductivity and pH of Water

1.3.1 Electrical Conduction Mechanism

According to the theory of electronic band structure [47], water and ice are dielectrics with negligibly small macroscopic electronic conductivity. The gap between the valence band and the conduction band of about 5.1 eV [48] (see Fig. 1.13) prescribes water to be a good dielectric, such as, e.g., quartz (6.3 eV) or diamond (5.5 eV). However, when an external electric field is applied, even very pure water shows anomalously high (for a dielectric) electrical conductivity $\sigma_{dc} = 5.5{\cdot}10^{-8}$ S/cm, which was measured by Kohlrausch [8]. This value is several orders of magnitude higher than those for quartz and diamond. Thus, the analogy between water and quartz, which was made by Bernal and Fowler on the basis of X-rays diffraction, is not complete, in particular, for the electrodynamic properties. What structural features make water so conductive in comparison with other dielectrics (Fig. 1.14)?

von Grotthuss (1806) (Fig.1.15a) suggested [50] that the relay-race transfer of protons among water molecules (see Fig.1.15b) provides an effective mechanism for charge transfer through the water sample. The Grotthuss mechanism allows one to describe quantitatively the anomalously high electrical conduction of water. This mechanism, the modern understanding of which is shown in Fig.1.15c, was used to explain the higher mobility of H^+ (H_3O^+) and OH^- ions in water in comparison with

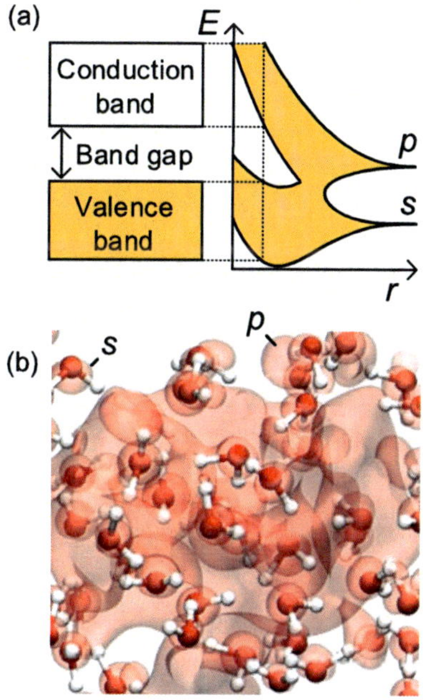

Fig. 1.13 **a** Energy band structure (left), and the energy levels as a function of the spacing between molecules (right). **b** Snapshots of bulk liquid water as observed by path integral molecular-dynamic simulations with isoprobability contours of the lowest unoccupied molecular orbital. Part (b) adapted from [49] with permission from Springer Nature

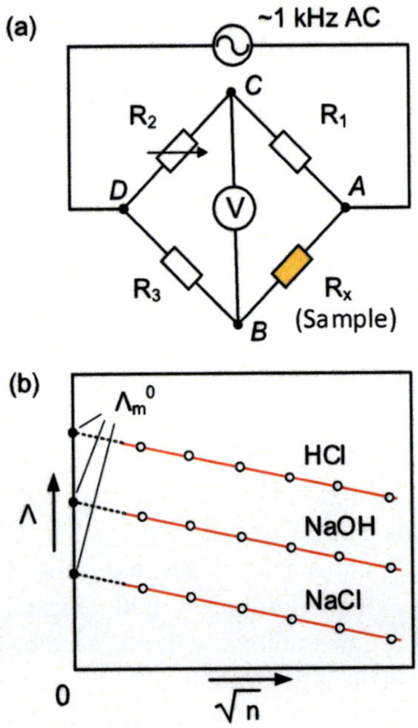

Fig. 1.14 **a** The autobalanced electric bridge circuit for the measurement of the electrical conductivity of water and aqueous electrolytes at low frequencies (typically around 1 kHz). The AC source is applied to the points A and D, the detector is applied between points B and C. **b** Molar electrical conductivity, Λ, of several aqueous electrolytes as a function of concentration. Λ_m^0 is the limiting molar conductivity at infinite dilution

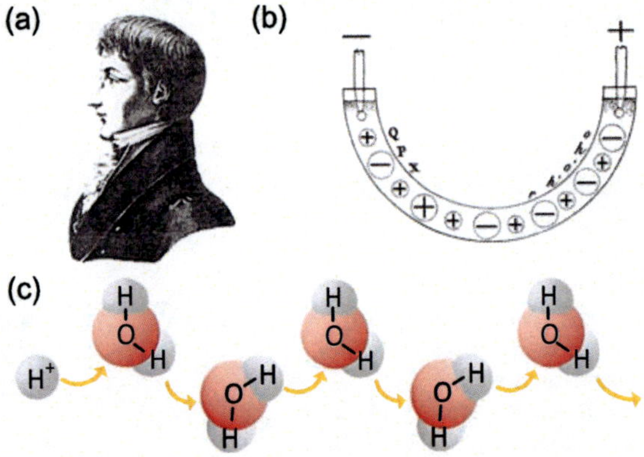

Fig. 1.15 **a** Theodor von Grotthuss (portrait by unknown artist, Bauska museum, Latvia). **b** His famous sketch of the electrical conduction mechanism in water, first published in 1806. **c** A modern representation of the Grotthuss mechanism of the proton transport

other ionic species [51, 55]. However, the details of intermolecular proton transport in water and its quantum-chemical quantitative parameters are still debated.

Since the charge carriers in water were identified, the question on how high their concentration is immediately arose. Using an autobalanced bridge shown in Fig. 1.14a, Kohlrausch [52, 53] showed that the conductivity of pure water does not depend on the frequency between 1 Hz and 10 kHz, and can be interpreted as the intrinsic conductivity of pure water.[12] Following the ideas of Arrhenius, Kohlrausch

[12]The exact value of the static conductivity depends on the type of electrodes, cell geometry, dissolved gases, and other factors, such as the measuring voltage, and can vary significantly.

assumed that H^+ (H_3O^+) and OH^- ions do not interact with each other,[13] and calculated the concentration, which corresponds to the observed conductivity level, using the following equation:

$$\sigma_{dc} = q \cdot n_{\pm} \cdot \mu, \tag{1.8}$$

where q and $n_{\pm}$ are the electric charge and concentration of carriers, respectively, and $\mu = \mathrm{q}{\cdot}D/k_BT$ is mobility, defined by the diffusion coefficient D of ions.[14] As one can see, the mobility and concentration in (1.8) are coupled (appearing as a multiplication). On the assumption of non-interacting ions, the mobility μ of the charge carriers can be found in experiments with aqueous electrolytes. For this reason, one can measure, for example, the molar conductivity $\Lambda = \sigma_{dc}/n$ of NaCl, NaOH, and HCl solutions at different concentrations, n, and evaluate the limiting conductivity at "infinite dilution," Λ^0_m, by extrapolating the experimental curves to low concentrations as shown in Fig. 1.14b. Assuming that ions of the solute do not interact with ions of solvent (water) at infinite dilution, and that the solute is completely dissociated, one can write the following equations:

$$\begin{cases} \Lambda^0_{\mathrm{NaCl}} = \Lambda^0_{\mathrm{Na}} + \Lambda^0_{\mathrm{Cl}}, \\ \Lambda^0_{\mathrm{NaOH}} = \Lambda^0_{\mathrm{Na}} + \Lambda^0_{\mathrm{OH}}, \\ \Lambda^0_{\mathrm{HCl}} = \Lambda^0_{H} + \Lambda^0_{\mathrm{Cl}}, \end{cases} \tag{1.9}$$

where Λ^0_{Na}, Λ^0_{Cl}, Λ^0_{H}, and Λ^0_{OH} are the limiting equivalent conductivities of the corresponding ions of the dissociation products of solute.

By solving the system (1.9), and applying the electro-neutrality principle,[15] one can obtain the concentration $n_{\pm}$ of H_3O^+ and OH^- ions, and then calculate their mobilities μ by (1.8). Table 1.1 shows the mobilities of different simple ionic species in water. The ionic mobilities of H_3O^+ and OH^- ions exceed the mobilities of other ions by several factors. One can also find that large ions have large mobility. The former fact was explained by the Grotthuss mechanism [50]. The latter was interpreted using solvation theory, which assumes that smaller ions have a large hydration shell (electrostatically attracted water molecules) [11].[16]

Although the method of infinite dilution allows one to obtain the concentration of charge carriers and calculate their mobilities, it is based on assumptions that ionic species are long-lived (do not participate in dissociation-recombine events and do not interact chemically with solvent), and that the concentration of intrinsic water

---

[13] The basis for this assumption was the fact that the conductivity of water in the specified frequency range decreased during its successive purification; however, no evidence that the charges in water do not interact with each other was presented.

[14] Equation (1.8) is known today as the Nernst–Einstein equation.

[15] The difference among concentrations of the ionic species of different signs is negligibly small.

[16] The author does not agree with this explanation, as all the listed cations have the same charge, and thus produce the same electrostatic force. There is no physical reason why some ions attract more water molecules than other. An alternative view has been proposed in [54] and is discussed in Chap. 5.

Table 1.1 The mobilities $\mu/(\times 10^{-8} m^2/s/V)$ of different ionic species in water and their ionic radii r_c/(pm), as obtained by the molar conductivity measurements at infinite dilution and X-ray crystallography [11]

Cations			Anions		
	μ	r_c		μ	r_c
H^+	36.2	0	OH^-	20.6	133
Li^+	4.01	90	F^-	5.74	119
Na^+	5.19	116	Cl^-	7.92	167
K^+	7.62	152	Br^-	8.09	182
Rb^+	8.06	166	I^-	7.96	206
Cs^+	8.00	181	–	–	–

ions is low enough to neglect their mutual interactions. Neither statement has been proved experimentally. Inasmuch as this model of electrolytes is widely used by the scientific community, there are still several open questions which need to be clarified. Why do different solutions have different molar conductivities at infinite dilution (as it is assumed that they should all be pure water)? Why do ions of the same charge attract different number of water molecules to form a solvation shell? Why do some ions, such as, e.g., Cs^+, have so-called negative hydration [11]? These questions cannot be answered using the Arrhenius theory of electrolytic dissociation and ion hydration.

1.3.2 *The Autoionization of Water*

The key element of electric current conduction by water is autoionization. Self-dissociation of water occurs when a water molecule, H_2O, donates the nucleus of one of its hydrogen atoms (i.e., a proton) to another water molecule, according to the formula:

$$H_2O + H_2O \leftrightarrow H_3O^+ + OH^-. \tag{1.10}$$

When this reaction goes in one direction, the donor molecule becomes a hydroxide ion, OH^-, and the acceptor molecule becomes a hydronium ion, H_3O^+. Note that the hydrogen nucleus, H^+, does not exist as an independent particle and forms hydronium ion with the nearest water molecule acceptor. That is why proton exchange occurs through a lower barrier than those required for the simple detachment of a proton from an isolated water molecule.

Figure 1.16 shows two basic intermediates of protonic transport in water. The four-coordinated (solvated) hydronium ion, or $H_9O_4^+$ (see Fig. 1.16a), is known as an Eigen cation. The intermediate state of the proton, between two molecules, or $H_5O_2^+$

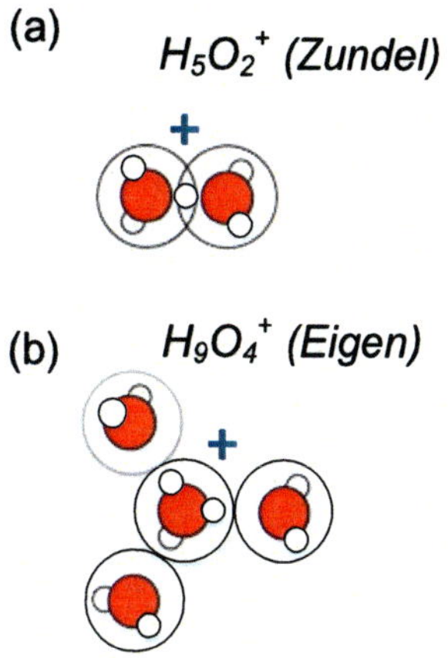

Fig. 1.16 The basic structural elements responsible for protonic current in water and ice. **a** Zundel cation, or a shared proton between two molecules with overlapped electronic clouds. **b** Eigen cation, or a stable four-coordinated state of the hydronium ion

(see Fig. 1.16b), is known as a Zundel cation [11]. The interconversion of these two cations provides the translational movement of protonic charge [56], and, as a result, electrical conductivity. Note that the oppositely charged OH^- species that compensate for the cations of water also participate in the ionic current. The mechanism of their migration is qualitatively similar to those for hydronium ions [57, 106]. One can simply imagine a "proton hole" instead of an "excess proton"ž. However, there are some quantitative differences in the potential functions of the corresponding Eigen and Zundel cations and anions [57], which are seen in their different mobilities (see Table 1.1).

Although both cations mentioned above represent two intermediate positions of a single mechanism of proton transport, the difference between them actually disappears when their dynamics are considered [58]. Figure 1.17 shows the free-energy profile of a Zundel cation. The proton-transfer energy barrier depends on the distance r_{OO} between the oxygen atoms of two neighboring species. When the separation between them is less than $r_{OO} \approx 2.36$ Å, the proton transfers without a barrier, depending only on the probability of the appropriate relative orientation of the molecules. The timescale, within which the two electronic clouds of the neighboring species are overlapped, does not exceed a fraction of a picosecond. In the context of the lifetime of the Eigen cation, which is a fraction of a picosecond [59], the formation of the Zundel cation can be considered as a collision between a hydronium ion and a nearby molecule. As a result of this collision, the proton transfer looks as if it just "sticks" from one molecule to another.[17] This important feature of protonic transport is missing in the Bernal–Fowler model.

Each excess proton/hole is a source of an electric field unscreened by electrons. Due to the specific structure of water molecules, described in Sect. 1.2.1, the molecules of the first solvation shell of the hydronium ion are oriented in a way

[17] As shown in Sect. 1.4, the mechanism, in which a proton sticks from one molecule to another without hopping, explains the fact that hydrogen and oxygen atoms have the same diffusion coefficients.

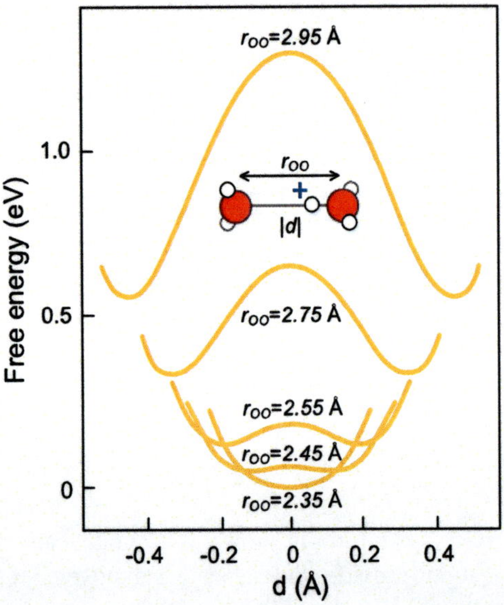

Fig. 1.17 The free energy of a Zundel ($H_5O_2^+$) cation as a function of distance d between oxygen atoms. Black dots show hydrogen atoms, open circles are oxygen atoms. The proton-transfer energy barrier disappears when the distance is 2.36 Å. Drawn using data from [60]

which favors effective proton transfer. The rate of proton transfer by the Grotthuss mechanism is limited by the rate of the adaptation of molecules around the hydronium ion [61], and thus it is diffusion controlled. Consequently, ionic transport in water is not just the electric field dragging solvated hydronium ions, but the rearrangement of water molecules in the way suitable for fast proton exchange, with a speed that is much higher than for conventional molecular diffusion [57]. However, the diffusion coefficient $D_{H+} = 9.3 \cdot 10^{-9}$ m^2/s [9, 62] is related to the charge of the proton, but not the mass of the proton, the diffusion coefficient of which is $D = 2.3 \cdot 10^{-9}$ m^2/s, the same as that for the water molecule (see Sect. 1.4). Water, therefore, has a structure in which the diffusion of mass and the charge does not coincide. It is in this property that water differs from other dielectrics and ionic conductors. Interestingly, the mobility of the proton charge in ice and water is of the same order of magnitude [63, 64]. This fact shows that the short-order molecular arrangement, which is the same for both ice and water, defines the proton conduction in both ice and water.[18]

1.3.3 *Pondus Hydrogenii (pH)*

The concentration of hydronium ions in pure water that has been established by DC electrical conductivity measurements was used as a reference value for the concept

[18]Note that the modeling of protonic transport by quantum-mechanical calculations is still difficult, because it involves too many particles to be calculated without simplifying assumptions [57].

of pondus hydrogenii (quantity of hydrogen), or simply pH.[19] The conductivity of pure water $\sigma_{dc} = 5.5 \cdot 10^{-8}$ S/cm used in (1.8) $n = [H^+] = 1/2 \cdot \sigma_{dc} k_B T/(q^2 D_p N_a)$, where $D_p = 9.3 \cdot 10^{-9}$ m^2/s is the proton self-diffusion coefficient [66], gives $n \approx 10^{-7}$ M, which is better known as pH = 7.

The concept of pH appeared as a result of the adaptation of the notion of the ionic product of water $K_w = [H^+] \times [OH^-]$, where $[H^+]$ and $[OH^-]$ are effective concentrations of hydronium and hydroxide ions, in a form that is more convenient for practical applications.[20] The ionic product is connected with pH by the formula pH = $1/2 \times pK_w$, where p is the decimal logarithm.

According to the definition [69], pH is a measure of activity of hydronium ions. When we assume pH = 7 for the pure water, it means that the *effective* concentration of hydroxyl ions is $n_\pm = 10^{-pH} = 10^{-7}$ mol/l, or $3 \cdot 10^{19}$ m^{-3}, or 10^{-7}%.[21] For an aqueous solution, pH values vary from 0 to 14, where pH = 0 corresponds to a very strong acid with seven orders of magnitude higher concentration of active H_3O^+ ions than in the pure water, and pH = 14 is for a very strong base with seven orders of magnitude higher concentration of active OH^- ions. Salts do not alter the pH value of water, as they do not bring any hydroxyl or hydronium ions to water, although they significantly increase its electrical conductivity [11] and decrease its dielectric constant (see Sect. 5.1 of Chap. 5 for explanation).

There is no direct way to determine the concentration of ions or the autoionization constant, pK_w. In order to detect the ions, one can use indicators, such as dyes or litmus paper, which change color depending on the activity[22] of hydronium ions and the intensity of the corresponding chemical reaction with water. Color indicators cover the whole pH range between 0 and 14 and, together with spectroscopic methods, allow a resolution better than 0.1 of the pH units. A more modern and precise method for pH measurements is based on the glass-electrode galvanic cell. This method measures the electromotive force between a reference electrode and an electrode sensitive to the hydronium ion activity when they are both immersed in the same aqueous solution [68]. Although this method is accurate, it requires calibration, and is very sensitive to impurities and experimental conditions.

The activity of hydronium ions can be also tested by nuclear magnetic resonance (NMR). The proton exchange broadens the line of the so-called chemical shift, which represents the resonant frequency of a nucleus relative to a standard in a magnetic

---

[19] In chemistry, pH (also known as "the potential of hydrogen" or "the power of hydrogen") is a scale used to specify the acidity or basicity of an aqueous solution. The pH scale is the negative of the base 10 logarithm of the "activity" of the H^+ ion.

[20] Interestingly, pH was originally developed in beer brewing [67] and apart from the modern definition had very pragmatic meaning.

[21] This value is several orders of magnitude lower than concentrations of such impurities in natural water as HDO, H_2O_2, CO_2, O_2, and N_2 in natural water, which are 10^{-2}, 10^{-7}, 10^{-4}, 10^{-4}, and 10^{-3}%, respectively.

[22] In chemical thermodynamics, activity a is a measure of the "effective concentration" of a species in a mixture, in the sense that the species' chemical potential depends on the activity of a real solution in the same way that it would depend on concentration for an ideal solution. The term "activity" in this sense was coined by Lewis in 1907.

field. The proton exchange causes additional confounding linewidth broadening of the NMR signal. One can measure the line width and compares it with the reference value for the liquid with a known concentration of ions. However, water normally serves as such a reference liquid, and another liquid with known ionic product is needed. Thus, the method is not absolute and requires some a priori information. If one assumes that proton exchange occurs via the Grotthuss mechanism, the line width reflects the concentration of hydronium ions. This method confirms that the high rate of proton exchange exists on a picosecond time interval [70]. However, such a short lifetime assumes a high concentration of hydronium ions, which are in dynamic equilibrium with neutral water molecules. This contradicts the conclusion made by Bernal and Fowler that the lifetime of H_2O is about 10 h [9], which assumes much higher (microsecond) ion lifetime. This fact shows one more inconsistency between Bernal–Fowler water and modern experimental data.

Some authors explain this inconsistency by the scrambling of protons between two neighboring molecules, which broadens the NMR line, but does not produce ionic species [70]. Quantum-chemical calculations, however, do not allow a high rate of proton scrambling, because of the high potential barrier and the low probability of the corresponding molecular orientations suitable for synchronized proton exchange [71]. Other authors [15] introduce the lifetime distribution of hydronium ions, and show that only those ions whose lifetime is long enough to be detected by relatively slow methods are pH active. In other words, most ions do not contribute to DC conductivity, but still are accountable by NMR. The latter model is in line with the fact that the pH of water shows an anomalously strong temperature dependence. This ionic model of water is described in Chap. 4. For instance, at 0 °C and 100 °C the pHs of pure water are 7.5 and 6.1, respectively. This complicates the application of the concept of pH for temperatures other than room temperature and requires accounting for the temperature dependence of the concentration of active ionic species in water (see Sect. 4.5.5 for details).

Figure 1.18 shows the temperature dependencies of the autoionization constant, pK_w, the DC electrical conductivity, σ_{dc}, and the molar conductivity, Λ_m, which was defined above. There are two "anomalies" that can be observed on these plots. First, the autoionization constant has a non-monotonic form and shows a minimum around 250 °C. Second, the difference in the concentrations of pH-active protons between this point and 0 °C exceeds four orders of magnitude, which is too high for the commonly accepted autoionization potential barrier of about 5.1 eV.

The origin of the strange behavior of the autoionization constant, pK_w, is described in Sect. 4.5.5, and lies in the different temperature dependencies of the equivalent conductivity, Λ_w^0, and the DC conductivity, σ_{dc}, of water, as one can see in Fig. 1.18. In order to understand this difference, one can rewrite the autodissociation constant in the following way:

$$K_w = [H_3O^+][OH^-] = [H_3O^+]^2 = (\alpha[H_2O])^2 = \\ = \left(\frac{\Lambda_w}{\Lambda_w^0}[H_2O]\right)^2 = \left(\frac{\sigma_{dc}}{\Lambda_w^0}\right)^2, \tag{1.11}$$

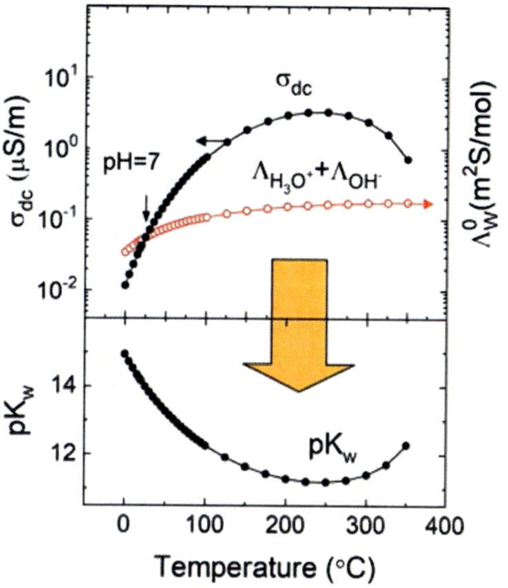

Fig. 1.18 The temperature dependencies of the electro-kinetic parameters of water: the specific conductivities, σ_{dc}; the equivalent conductivity, Λ_w^0; and the coefficient of the autoionization constant, pK_w, derived from them. The dotted lines mark T = 25 °C, the temperature at which the concept of pH is determined and $pK_w = 14$, pH = 7. Experimental data from [72]

Table 1.2 Parameters of (1.11) for room-temperature water

n_{H_2O} (M)	Λ_w (S· m²/mol)	$\Lambda_{H_3O^+}$ (S · m²/mol)	Λ_{OH^-} (S · m²/mol)	Λ_w^0 (S · m²/mol)
55.5	$9.90 \cdot 10^{-7}$	349.8	198.5	548.3
pH	σ_{dc} (S/cm)	α	$n_{H_3O^+}$ (M)	pK_w
7	$5.5 \cdot 10^{-8}$	$1.8 \cdot 10^{-9}$	$1.0 \cdot 10^{-7}$	14

where $\alpha = \Lambda_w/\Lambda_w^0$ is the degree of H_2O dissociation; $\Lambda_w = \sigma_{dc}/n_{H_2O}$ is the molar conductivity; and $\Lambda_w^0 = \Lambda^0_{H_3O^+} + \Lambda^0_{OH^-}$ is the equivalent conductivity, defined by the sum of the mobilities of H_3O^+ and OH^- ions as obtained from experiments with infinite dilution. Table 1.2 shows different quantitative estimates of the parameters of (1.11) at room temperature as the concept of pH is adjusted for room temperature (see black arrow in Fig. 1.18). However, the ratio represented by the last term in (1.11) is not conserved when the temperature changes. The DC conductivity values change faster than those for Λ_w^0 (see Fig. 1.18) which leads (see yellow arrow) to the anomalous temperature behavior of pK_w, and, as a consequence, pH.

Summarizing, the concept of pH is based on the poorly defined notion of the activity of hydronium ions in water and aqueous solutions. It reflects the diffusion-averaged concentration of ions, and does not account for the fluctuation-borne short-lived ions, the concentration of which is much higher [15, 73] in comparison with the long-lived ions detected by glass electrode, or conductometric techniques (see Chaps. 3 and 4). However, the short lifetime makes them "invisible" to static conductivity measurements. Let us consider the following abstract analogy. Figure 1.19 shows the photo of the Paris boulevard at the midday taken in 1838.[23] It looks empty,

[23]This photo is the first known photo of a human.

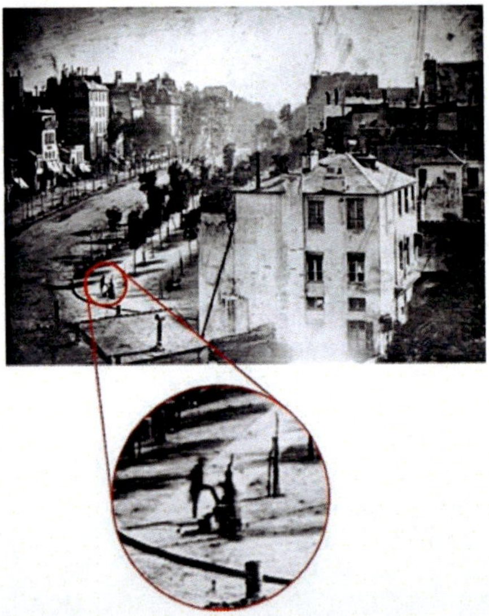

Fig. 1.19 The photograph of Boulevard du Temple in Paris by Louis Daguerre (1838). The time averaging makes people invisible, except those who stay at the same state (see inset). See the text for an analogy with the averaged dynamics of protons in water. Open-source picture

because moving objects "dissolved" during the exposure, which at that time was very long (about 20 min). Only one person (see inset) is detected, who was standing in the same place during the whole exposure. By analogy, only those protons "visible" at observation times that are long-lived enough to produce a signal for the period of the sensing electric field. Further analysis of the concept of pH in new ionic structural models of water is given in Sect. 4.5.5.

1.4 Self-diffusion by Isotopic Tracers

The study of water molecule self-diffusion provides a valuable piece of information for understanding its transport properties. Due to the indistinguishability of atoms and molecules by conventional methods, the self-diffusion coefficient of water was poorly studied until isotopic indicators were used for this purpose in the second half of the twentieth century, which became very affordable due to the beginning of the nuclear era. Isotopic tracers are molecules which behave as any other molecule in the substance being studied, but have some distinguishing property, such as a different mass or radioactivity, by which it can be detected and identified among the native molecules of the substance.

Isotopic substitution is a more direct way to obtain the self-diffusion coefficient than nuclear magnetic resonance [65] and conductometry, discussed above. This method works as follows. Before diffusion the capillary (1) (see Fig. 1.20) is filled with a tracer solution of known concentration. Then it is held vertically in a large circulating bath of pure ordinary water of the same temperature. After diffusion the

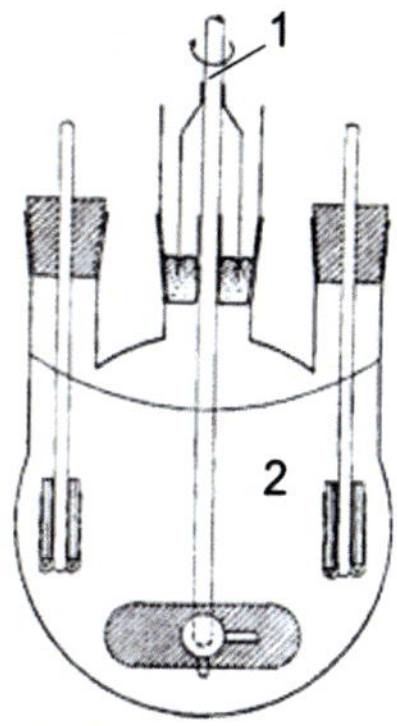

Fig. 1.20 Capillary apparatus for measuring self-diffusion coefficients of liquids: (1) a capillary with an isotope tracer, (2) a vessel with ordinary liquid. The arrow shows the direction of rotation. Reprinted with permission from [75] Copyright 1951 American Chemical Society

capillary is taken out of bath, the solution is well mixed, and the residual concentration is determined by weight. The diffusion coefficient, D, is calculated by

$$\frac{Dt}{l^2} = \frac{4}{\pi^2} \ln\left(\frac{8}{\pi^2} \times \frac{c_0}{c_{av}}\right), \tag{1.12}$$

where c_{av} is the average concentration of the tracer in the capillary at time t, c_0 is the initial concentration of the tracer, and l is the length of capillary. Using different isotopes, such as H_2O^{18}, HTO^{16}, and H_2O^{16}, and measuring the residual concentration as a function of time using a sensitive balance, one can distinguish the diffusion coefficient for hydrogen and oxygen atoms. The ability to obtain atomic diffusion coefficients separately by using "marked" oxygen or hydrogen atoms is a distinctive feature of the method of isotopic substitution.

This method is also applicable for ice, but the experimental procedure is different [74]. The isotopic tracer, e.g., tritiated water, is evaporated on the flat surface of a cylindrical ice sample. Since tritium emits β particles, the surface activity can be measured by a gas flow counter, which shows the residual concentration of tritium in the surface as a function of time. After a certain time, the sample is sliced, and the surface activity is measured as a function of the distance from the initial surface along the cylinder. Finally, the diffusion coefficient D is reconstructed using Fick's diffusion equation.

Figure 1.21 shows the experimental temperature dependencies of the diffusion coefficients of different isotopic tracers obtained by the above techniques. The molecules with a deuterium atom (D $\equiv {}^2$H), a tritium atom (T $\equiv {}^3$H), or an oxygen isotope (O^{18}) show the same, within the experimental uncertainty, diffusion coefficients with the equivalent activation energies. Data are independently confirmed by the modern NMR data, which gives the same values of H_2O-molecule diffusion, and additionally extends the temperature range. Figure 1.21 shows the averaged fit of the data points, and the best-fit parameters for all curves are given in Table 1.3. Both the diffusion coefficients and their activation energies coincide within the margin of errors for all studied molecular species. In other words, hydrogen and oxygen atoms

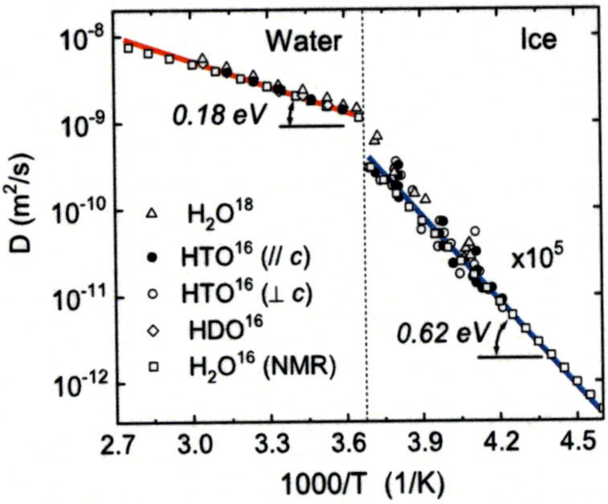

Fig. 1.21 The diffusion coefficients of isotopes in **a** liquid water, and **b** ice versus reciprocal temperature, as obtained by the isotopic tracer diffusion and NMR. Experimental data are collected from [63–65]. Numbers with eV show the activation energies, also given in Table 1.3

Table 1.3 The parameters of the temperature dependence of the diffusion coefficients of isotopes and ions in water and ice in accordance with the formula $D = D_0\exp(-E_a/k_BT)$, where D_0 is the pre-exponential factor and E_a is the activation energy.

Isotope/Ion	Water		Ice	
	D_0 (m²/s)	E_a (eV)	D_0 (m²/s)	E_a (eV)
H_2O^{16}	$2.3{\cdot}10^{-6}$	0.177±0.003	$1.1{\cdot}10^{-3}$	0.619±0.001
H_2O^{18}	$4.3{\cdot}10^{-6}$	0.188±0.005	$1.2{\cdot}10^{-1}$	0.709±0.038
HDO^{16}	$5.4{\cdot}10^{-6}$	0.198±0.008	No data	No data
HTO^{16}	$4.2{\cdot}10^{-6}$	0.192±0.001	$4.2{\cdot}10^{-6}$	0.628±0.042
H_3O^+/OH^-	$5.0{\cdot}10^{-4}$	0.174±0.003	No data	No data

have the same diffusion coefficient as H_2O molecules in both ice and water. The molecular self-diffusion coefficients of water and ice near the melting point are $D_W = 2.0{\cdot}10^{-9}$ and $D_I = 1.0{\cdot}10^{-14}$ m²/s, respectively, showing a difference of five orders of magnitude. The activation energy of molecular diffusion for water and ice differs by a factor of 3 (see Table 1.3). Both the latter facts confirm the higher cooperativity of ice particles in comparison with those in water.

The same diffusion coefficient for the oxygen and hydrogen atoms of water and ice with the equivalent activation energies has been previously discussed by Eisenberg and Kautzmann [3]. They conclude that oxygen and hydrogen atoms, once assembled into H_2O molecules, diffuse together. In other words, no reconstructions of the molecules with time were expected. This assumption contradicts many modern experimental data. As we have seen above, the hydrogen atom shows a higher diffusion coefficient, $D_p = 9.3{\cdot}10^{-9}$ m²/s, as a result of the molar conductivity analysis (see Sect. 1.3.1), showing a five times higher diffusion coefficient than that observed by the isotopic tracer technique. The NMR broadening line also assumes intense proton exchange between the molecules (see above).

Figure 1.22 compares the isotopic (D_w) and conductometric (D_p) diffusion coefficients, which are plotted as functions of temperature and pressure. The latter is

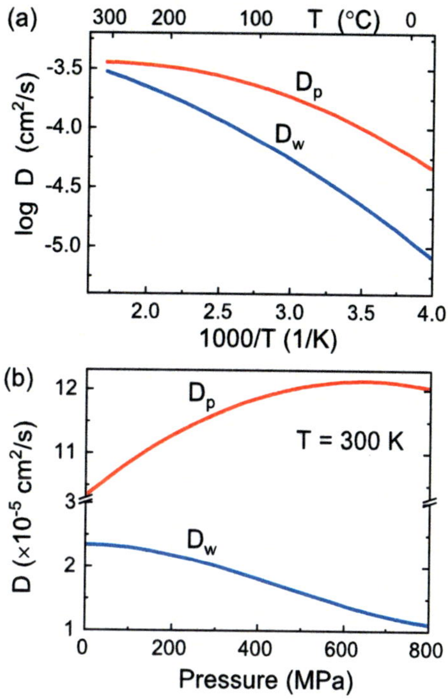

Fig. 1.22 The diffusion coefficient of a proton, D_p, obtained by the equivalent conductivity measurement (see Sect. 1.3.1) and the self-diffusion coefficient of H_2O molecules (and protons), D_w, measured by the isotopic substitution technique: **a** as a function of temperature; **b** as a function of pressure. Data from [76]

calculated using (1.8), where $\mu = D_p/k_B T$, while the former is measured directly. One can see that these coefficients do not coincide. As the pressure increases, the diffusion coefficients first diverge, but then converge again. Thus, there is a contradiction between the concepts of the Grotthuss mechanism for proton diffusion and the diffusion data presented here. In the experiments with labeled atoms, protons do not show increased mobility, which is expected by the relay-race mechanism of proton hopping. The latter implies that at the each step of proton transfer, it will gain an additional spatial displacement in comparison with an oxygen atom, and, as a result, should have a higher diffusion coefficient according to the following formula:

$$D = \frac{\Delta r^2}{6t}, \tag{1.13}$$

where Δr is the first step of diffusion, which for proton hoping should be δr longer, thus giving a higher diffusion coefficient D. But, the increased mobility of protons was not observed with isotopic tracers.

The source of this discrepancy is in the incorrectness of the comparisons of D_w and D_p, because they were obtained using different methods. The first coefficient refers to the mass, but the second is related to charge. If charge can move in the relay-race manner, it moves ahead of the specific proton, because only one of three protons of an H_3O^+ ion can be transferred by the Grotthuss mechanism: the transferring proton is not necessarily the same one which brought this charge from the previous step of the diffusion.

Thus, the charge transfer is carried out through a proton "sticking" from H_3O^+ to H_2O, as shown in Fig. 1.16a. In this way, $\delta r = 0$. That is why a hydronium ion, or proton, always has the same diffusion coefficient as an oxygen atom, because the proton does not jump, but sticks from one molecular species to another, just like a relay race, where the baton has the same mobility as the runner. This mechanism assumes a finite lifetime of H_2O molecules, which varies from 10 h [77, 78] to ms [79], depending on the initial assumptions and the model.

1.5 Diffusion by Neutron Scattering

Quasi-elastic neutron scattering is a relatively new technique [80]. The first accurate data on water appeared in the 1970s when Powles and Page measure the mixtures of heavy and light water [81, 82]. But, the first results on water were even earlier; Singwi in the 1960s suggested using neutrons for the study of the diffusive motion of water molecules [83]. When neutrons interact with hydrogen atoms, there could be an exchange of momentum and energy, which depends on the diffusive motion of the water molecules. Since neutrons interact with protons only, we can track hydrogen atoms, but cannot say anything about diffusion of oxygen atoms. In this context, neutron scattering complements X-ray diffraction (see Sect. 1.2). Note, however, that as we saw above, hydrogen and oxygen atoms are not necessarily attached to each other, even if they have the same diffusion coefficient (see Sect. 1.4).

Figure 1.23 shows a typical quasi-elastic neutron scattering spectra for water. Scattering intensity is plotted as a function of the energy transfer ΔE. The natural line of water appears as a result of the subtraction of an experimentally measured curve by the spectrometer resolution function. Figure 1.24 shows the half-width of the corresponding natural quasi-elastic scattering line for water as a function of wave vector κ. This curve contains all the information on the diffusion of protons. But, in order to extract this information, one again needs an initial assumption on atomic-molecular dynamics, and a preliminary understanding of the type (rotations, translations, drift, etc.) of molecular motion in water.

The Bernal–Fowler model of water (see Sect. 1.3.1) assumes that diffusion occurs by an interstitial mechanism through the large displacements of molecules from a stationary vibrational state to another quasi-equilibrium state. This idea continues in modern quantum-chemical calculations, which show that molecular diffusion in water goes through so-called wait-and-switch dynamics [85]. The approach assumes that water molecules have an oscillatory-diffusional motion with a diffusion coefficient D of $\approx 10^{-9}$ m^2/s, periodically lingering for a time $t_0 \approx 4$ ps in localized vibrational states. The main difference for the Bernal–Fowler approach is that the oscillatory state is not stationary, because the center of oscillations drifts with time, obeying Brownian motion. Figure 1.25a sketches the trajectory of a proton that corre-

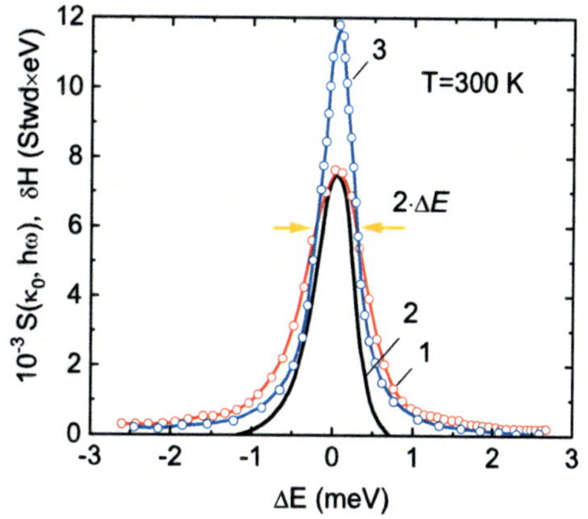

Fig. 1.23 Lines of quasi-elastic neutron scattering for water at $\kappa_0 = 1.21$ Å^{-1}, T = 300 K: 1 is the experimental curve, 2 is the spectrometer resolution function, and 3 is the natural line of water. Data from [84]

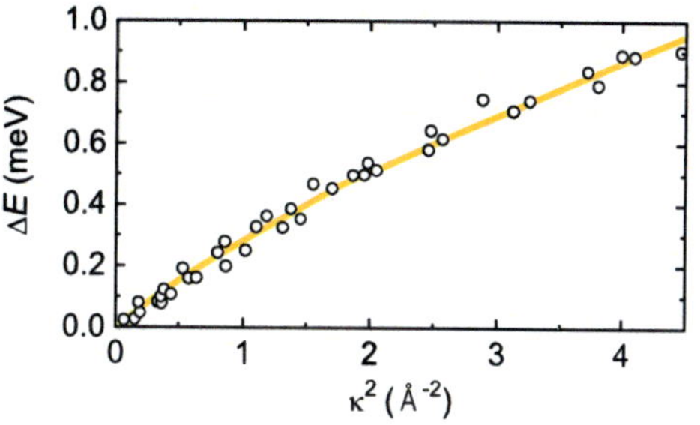

Fig. 1.24 The half-width of the natural quasi-elastic scattering line for water as a function of the squared wave vector κ^2 at T = 300 K. Data from [84]

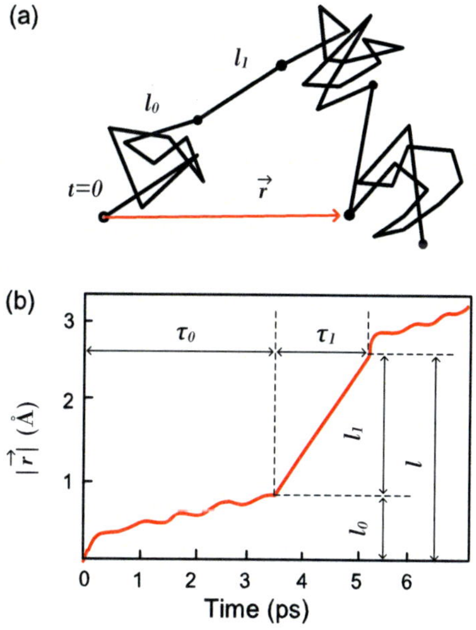

Fig. 1.25 The spatial-time dynamics of protons in water: **a** The proton trajectory and **b** the mean square displacement according to quasi-elastic neutron scattering data within the wait-and-switch model. For explanation, see text

sponds to the model of diffusion described above (see Sect. 1.4), where **r** is the radius vector. Figure 1.25b shows the mean square displacement, one period of which has two components. The first one, l_0, corresponds to the drift of the equilibrium point of the oscillatory state within time τ_0. The second one, l_1, is the proton displacement to a new equilibrium position within time τ_1. For the observation times where $t \gg \tau_0 + \tau_1$, the diffusion coefficient is equal to the averaged coefficient D. For $t \sim \tau_0$ or τ_1, the diffusion coefficient becomes time dependent, which, at the first glance, can be interpreted as anomalous diffusion,[24] but which reduces to the sum of the normal diffusion processes on closer consideration.

Neutron scattering data allow one to obtain the characteristic diffusion lengths and times shown in Fig. 1.25. The algorithm for the determination of these diffusion parameters was suggested by Singwi. The differential scattering cross section is determined by [83]

$$\frac{d^2\sigma}{d\Omega d\omega} = \frac{\kappa}{\kappa_0}\left(b_c^2 S(q,\omega) + b_i^2 S_S(q,\omega)\right), \tag{1.14}$$

where κ and κ_0 are the wave vectors of scattered and incident neutrons, respectively; b_c is the coherent and b_i is the incoherent scattering amplitudes; and $E = \hbar\omega$ and $q = \hbar(\kappa\text{-}\kappa_0)$ are the energy and angular momentum of the scattering process, respectively. The general scattering function $S(q, \omega)$ is the Fourier transform of the space-time correlation function $G(r, t)$, and $S_s(q, \omega)$ is the Fourier transform of the autocorrelation function $G_s(r, t)$. The latter determines the probability of finding a particle at the point (r, t), if at time $t = 0$ the same particle was at the origin.

At the classical limit, and assuming that the translational and rotational degrees of freedom are not correlated, the interpretation of the $G(r, t)$ function is based on the idea that an isolated hydrogen atom, being at the origin at the time $t = 0$, determines the probability of finding an atom in a unit volume at point r at time t. The atom in position (r, t) can be either the same that was at the origin at the time $t = 0$, or a different hydrogen atom.

$G_s(r, t)$ is defined by the sum:

$$G_s(r,t) = \sum_i F_i(r,t), \tag{1.15}$$

where F_i is given by

[24] Anomalous diffusion is a diffusion process with a non-linear relationship between the mean squared displacement (MSD) and time.

$$F_0(r,t) = g(r,t)p(t),$$

$$F_1(r,t) = -\int_0^t dt_1 \int_0^{t_1} dr_1 q(t-t_1)h(r-r_1, t-t_1)p'(t_1)g(r_1, t_1),$$

$$F_2(r,t) = (-1)^2 \int_0^t dt_2 \int_0^{t_2} dt_1 \int dr_2 dr_1 p(t-t_2)g(r-r_2, t-t_2)q'(t_2-t_1)\times$$
$$\times\, h(r_2-r_1, t_2-t_1)p'(t_1)g(r_1, t_1),$$

$$F_{2n}(r,t) = (-1)^{2n} \int_0^t dt_{2n} \int_0^{t_{2n}} dt_{2n-1} \cdots \int_0^{t_2} dt_1 \int \cdots \int dr_{2n} dr_{2n-1} \cdots dr_1 p(t-t_{2n})\times$$
$$\times\, g(r-r_{2n}, t-t_{2n})q'(t_{2n}-t_{2n-1})h(r_{2n}-r_{2n-1}, t_{2n}-t_{2n-1}) \cdots p'(t_1)g(r_1, t_1). \tag{1.16}$$

Formula (1.16) shows that the problem of determining $G_s(r, t)$ reduces to the definition of the following quantities:

- The probabilities $g(r, t)$ and $h(r, t)$ of finding a proton in position r at time t starting from the origin at time $t = 0$, when it performs oscillatory motion around the equilibrium position and diffusion motion between two equilibrium positions:

$$g(r,t) = \left[2\pi\gamma(t)\right]^{-3/2} \exp(-r^2/2\gamma(t)), \tag{1.17}$$
$$h(r,t) = [4\pi D_1 t]^{-3/2} \exp(-r^2/4D_1 t), \tag{1.18}$$

  where $\gamma(t)$ is a half-width of the autocorrelation function.
- The probabilities $p(t) = \exp(-t/t_0)$ and $q(t) = \exp(-t/t_1)$ that the proton will remain in the same state of vibrational/diffusional motion for a longer time t.
- The probabilities $p'(t) = p(t') - p(t)$ and $q'(t) = q(t') - q(t)$ that the proton will leave its vibrational/diffusional state during the time $t - t'$, switching to a diffusional/vibrational state.

Singwi and Sjölander assumed [83] that the translational diffusion coefficient within time τ_1 is determined by the Einstein–Smoluchowski formula:

$$D_1 = \frac{l^2}{6\tau_1}, \tag{1.19}$$

where l is the displacement of the particle within the time of the diffusional dynamics.

Later Oskotsky introduced [86] the continuous diffusion model of molecules in an oscillatory state and showed that the total diffusion coefficient, D, consists of the coefficients of continuous, D_0, and jump-like, D_1, diffusion:

$$D = \frac{d^2 + l^2}{6(\tau_0 + \tau_1)}, \tag{1.20}$$

Table 1.4 Proton diffusion parameters using neutron scattering data, defined by (1.22). D_0, D_1, and D are the local-continuous, discontinuous, and general diffusion coefficients, respectively. l_0, l_1, and l are the lengths of vibrations, of hops, and between neighboring oxygen atoms, respectively. τ_0 and τ_1 are the waiting and hopping times, respectively. See also Fig. 1.25

D_0 (m²/s)	D_1 (m²/s)	D (m²/s)	l_0 (Å)	l_1 (Å)	l (Å)	τ_0 (ps)	τ_1 (ps)
$0.7 \cdot 10^{-9}$	$5.8 \cdot 10^{-9}$	$2.3 \cdot 10^{-9}$	0.31	1.6	2.8	3.7	1.7

where d is the displacement of the proton during the time of the complete elaboration of the "heat cloud" during oscillatory motion. This model is better suited for the trajectory shown in Fig. 1.25. If $d^2 \ll l^2$, (1.20) gives the relation between the coefficients D_1 and D in the form:

$$D_0\tau_0 + D_1\tau_1 = D(\tau_0 + \tau_1), \tag{1.21}$$

showing that the averaged mean square displacement is an additive sum of drift and hopping. If $\tau_1 \ll \tau_0$, the shape of the quasi-elastic scattering peak has a Lorentzian shape with a width at half maximum of

$$\Delta E = \frac{2\hbar}{\tau_0}\left(1 + \kappa^2 D_0\tau_0 - \frac{e^{-2W}}{1 + \kappa^2 D\tau_0}\right), \tag{1.22}$$

where $2W = \kappa^2 D\tau_0(d^2/l^2)$ is the Debye–Waller factor. Figure 1.24 shows that (1.22) perfectly fits the experimental data on neutron scattering. The best-fit parameters are given in Table 1.4.

Table 1.4 shows that neutron scattering data assume two characteristic times of diffusion, which correspond to vibrational or translational motion. The elementary period of diffusion is $t = t_0 + t_1 \approx 5$ ps. The corresponding distance in which a proton moves for time t is equal to 3.1 Å, and consists of two parts: $l_0 \approx 0.3$ Å and $l_1 \approx 2.8$ Å (see Table 1.4 and Fig. 1.25). The latter coincide with the diameter of a water molecule, or with the average distance between two neighboring molecules. As long as the probability that all molecules can rotate on the same angle of 180° is very small, one can assume the distance l_1 corresponds to the intermolecular transfer of a proton. This is the same as that observed in the NMR spectroscopy of water (see Sect. 1.3.3). The time necessary for such a transfer is determined by the tracer diffusion coefficient $D_{tr} = 2.3 \cdot 10^{-9}$, because proton transfer is limited by self-diffusion (see Sect. 1.5). $D_{tr} = D$, as follows from the Table 1.4, which indirectly confirms the assumption that neutron scattering line width is determined by intermolecular proton exchange, rather than molecular reorientations, although the latter help molecules to adjust the position that is suitable for proton transfer.

Neutron scattering shows that the protons of water oscillate around a drifting equilibrium position within $t_0 \approx 4$ ps, and then switch to another equilibrium position within $t_1 \approx 2$ ps, showing a diffusion-oscillatory motion. Note that the vibrational state residence time, t_0, is several times longer than the period of oscillations. After several collisions, the proton is transferred to a new position within time t_1, after which it again appears in the new oscillatory state. Then the process repeats. Protons, when in oscillatory states, still undergo continuous diffusion with the center of oscillation in a quasi-equilibrium vibrational state at the same time.

1.6 Water in Molecular-Dynamic Simulations

1.6.1 Methods of Water Modeling

Methods of computer simulation, such as molecular dynamics (MD), molecular mechanics (MM), dissipative particle dynamics (DPD), and Monte Carlo (MC), have been actively used for the studies of water and ice since the pioneering works of Barker and Watts, and Rahman and Stillinger [87, 88]. Nowadays, computer simulations are an important part of studies of water and ice. Many experimental data on water, including electrodynamic parameters, were verified by computer simulations.

The main problem of molecular-dynamic studies is to find the potential U of intra- and intermolecular interactions, which would provide a description of the experimentally measured parameters of water, e.g., structural, such as density, viscosity, surface tension; electrical, such as dielectric constant, infrared spectrum, electrical conductivity; and thermodynamic, such as heat capacity and thermal conductivity. Figure 1.26 shows the landscape of the main modern methods of computer simulation, which are used to study the microscopic structure of water and ice. One can see that different methods are applicable for different time intervals, and can also account for the different number of particles (molecules or atoms). The MC and MD methods cover the largest range of times (including diffusion and structural relaxation), but require knowledge of potentials, which is hard to assume a priori. That is why there are huge variety of three-, four-, and five-point models of water molecules, which are used to model different water properties. The most popular models are SPC/E, TIP4P, TIP4P/Ice, and TIP5P [89]. In addition, some simple semi-classical approaches have recently come under the spotlight again [90]. The wide range of models is a consequence of the fact that different properties are described by different potentials. To date, there is no model that can quantitatively describe all the properties of water, and the search for a "universal" model continues.

The MD approach is based on the idea that water is an ensemble of H_2O molecules. The first molecular models were built on the structure proposed by Bernal and Fowler (see Sect. 1.3.1 and Fig. 1.2), the model that has been around since the pre-computer era [91]. When first computers appeared, it was immediately shown that rigid non-polarizable models fail to reproduce the basic properties of water, including electrodynamic parameters (see, for example, [92]). In order to improve the model, the potentials that take into account the polarizability of H_2O molecules due to the influence of the environment have been suggested [93, 94]. In some calculations, the molecular dipole moment μ increases by 60% compared to those observed in the gas phase [95]. These changes to the molecular structure are usually modeled by

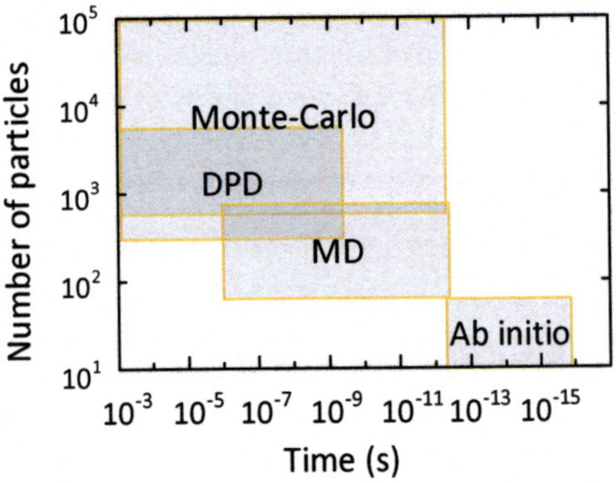

Fig. 1.26 Computer simulation methods by the number of particles and the observation time of the corresponding microscopic processes. DPD is dissipative particle dynamics. MD is molecular dynamics

adding oscillating, distributed, or variable charges, and require some additional computational cost. However, there are experiments in which such overpolarized H_2O molecules were not confirmed [96, 97]: in infrared spectra, the O–H stretch frequency changes insignificantly compared to the gas phase, and neutron diffraction does not confirm significant perturbations of the hydrogen atom positions. Thus, although certain successes in the modeling of the water properties by MD methods have been unambiguously achieved, there is still a question on how close these models are to reality [98].

The growth of computer efficiency by the mid-1990s made it possible to implement quantum chemistry methods for the calculation of the electronic structure of water molecules. Nowadays, so-called ab initio methods are widely used for the simulation of water. This approach is based on solving the Schrödinger equation in the Born–Oppenheimer approximation,[25] and yield electron densities, energies, and other properties of the molecular water system. Although the method does not use any input parameters except initial atomic coordinates, it requires high computational costs that makes the calculation of large molecular systems difficult. That is why the pure ab initio method is not applicable for real systems, and deals with only a few molecules (see Fig. 1.26).

The density-functional theory (DFT) is typically used to minimize the computational costs of ab initio calculations. In hybrid ab initio-DFT simulations, the electronic structure is evaluated using the electrostatic potential of the electrons for atoms of the molecular system. The DFT potential is constructed as the sum of external potentials U_{ext}, which can be determined by the X-ray spectroscopy of water vapor, and an effective potential U_{eff}, which represents interelectronic interactions and is generally unknown. Using this theory, the properties of water can be determined by

[25]The assumption that the motion of atomic nuclei and electrons in a molecule can be treated separately, as the nuclei are much heavier than the electrons.

using so-called functionals.[26] Although the approach is very useful for the study of the structure that corresponds to the potential energy minimum, the DFT fails to describe the intermolecular interactions of water, account for intermolecular charge transfer, the transition states of protons and proton holes, or find any far-from-equilibrium potential energy surface [99]. Moreover, with current computers, a sufficiently accurate simulation is possible only for processes shorter than a few picoseconds [100] (see Fig. 1.26). Nevertheless, the DFT method continues to provide useful information on the local structure of water and ice, including nuclear quantum effects [101], which are beyond the consideration in conventional MD-potential models, but seems crucial for the electrodynamics of water (see Chap. 4).

The most popular among ab initio approaches which have been successfully applied to water and ice is Car–Parrinello molecular dynamics [102]. This is based on the symbiosis of DFT and Kohn–Sham functions [91]. This method does not require any potentials in an analytical form. Instead, wave functions are specified. The parameters of water molecules in neutral [95, 103, 104], protonated [105], and deprotonated [106, 107] states, as well as in ice [108] have been carefully modeled by the Car–Parrinello method for the past several decades. Many atomic-molecular mechanisms have been modeled by introducing adjustable parameters. However, the universal parameters of interacting water molecules which can describe all the properties of water have not yet been found. There are some interesting examples below, which are related to the electrodynamic properties of ice and water.

1.6.2 The Simulation of the Electrodynamic Parameters of Water and Ice

Geissler et al. [77] analyzed the autodissociation of water by sampling ab initio molecular-dynamic trajectories. Identifying the rare fluctuations in solvation energies, the authors studied the transfer of protons between molecules. This is an important process for the electrodynamics of water. Calculations showed that the proton "wire" (the chain of proton-transfer events) reaches a length of at least three molecular diameters. This result coincides with a previous study [109], and additionally shows that if the proton wire remains unbroken, the ions recombine rapidly within 100 fs.[27] This time interval was later confirmed by Marx [50]. In accounting for the concentration of ionic species that appear/disappear due to autoionization events, Geissler et al. rely on the classic paper of Eigen who states, on the basis of the early conductivity measurements (see Sect. 1.3.1), that the average H_2O molecule lifetime

[26]Hybrid functionals are a class of approximations of the exchange-correlation energy functional in DFT which incorporate a portion of the exact exchange from Hartree–Fock theory with the rest of the exchange-correlation energy from other sources (ab initio or empirical).

[27]Note that this time corresponds to the so-called OH-band (3,600 cm^{-1}) of the infrared spectrum of liquid water.

is about 11 h [110].[28] In order to satisfy this huge lifetime, Geissler and co-authors concluded that most water ions recombine quickly, without contributing to DC conductivity; however, some of them can survive for a long time, and the transient ionic species may provide an experimentally detectable electrodynamic signal.[29]

Hassanali et al. [73] analyzed the recombination of H_3O^+ and OH^- ions in water, taking into account nuclear quantum effects: proton tunneling and its zero point energy. They found that the recombination of excess proton and proton hole occurs in two stages: the compression of the chain of molecules connecting the corresponding ions in about 0.5 ps, and the neutralization of the compressed chain in 65 fs. Thus, it was established that recombination requires the coordinated motion of several molecules, which makes this process unlikely in real temperature conditions. However, one should note that phonons can provide the necessary conditions for coordinated molecular dynamics even at room temperature.

Rick and Haymet [111] studied the dielectric properties of ice Ih, using a Monte Carlo algorithm for sampling over proton configurations. They used non-polarizable (SPC/E and TIP4P) and polarizable (TIP4P-FQ) molecular models to calculate the static dielectric constant, $\epsilon(0)$, of ice. A good agreement with experimental data was obtained only in the temperature range 150–270 K, and only for the polarizable model TIP4P-FQ, which requires a phenomenological g-factor that was adjusted to reach the required value of $\epsilon(0)$. The physical meaning of the *g*-factor, which allowed the authors to fit the experimental data, is unclear, and it can be regarded as a phenomenological parameter, the physical meaning of which is missing. For further details, see Sect. 2.4.

Aragones et al. [112], continuing the previously mentioned study, calculated the dielectric constants of ices Ih, III, V, VI, and VII for several water models using the Monte Carlo method and an approximation in which proton-disordered configurations satisfy the Bernal–Fowler ice rules. Figure 1.27 shows that the SPC/E, TIP5P, TIP4P, TIP4P/2005, and TIP4P/ice models of water are unable to reproduce simultaneously the experimental dielectric constants of water and ice Ih. The authors state that the predictions of the Bernal–Fowler model for the dielectric constant differ, in general, from that obtained rigorously by computer simulations because proton-disordered configurations, which satisfy the Bernal–Fowler ice rules, differ in their energies. They concluded that non-polarizable models cannot describe the dielectric constants of different condensed phases of water because their dipole moments (about 2.3 D) are much smaller than those required for the fitting of the experimental data, and, as estimated by the first-principles calculations, should be about 4.0 D. The predictions of TIP4P models provide an overall qualitatively correct description of the dielectric constant of the condensed phases of water only when the dipole moment of the model is scaled to the estimated value obtained from first-principle calculations

[28]In fact, Eigen uses the adjective "apparent" for the 11-hour lifetime, and highlights that the "true" lifetime of H_2O molecules is determined by the intensity of proton exchange and lies in the microsecond time interval, but, because of the indistinguishably of protons, cannot be detected. Latter, the NMR method confirmed the short lifetime of water molecules (see Sect. 1.3).

[29]These short-lived ionic species of water have been recently experimentally detected in the infrared spectra of light and heavy water mixtures [32].

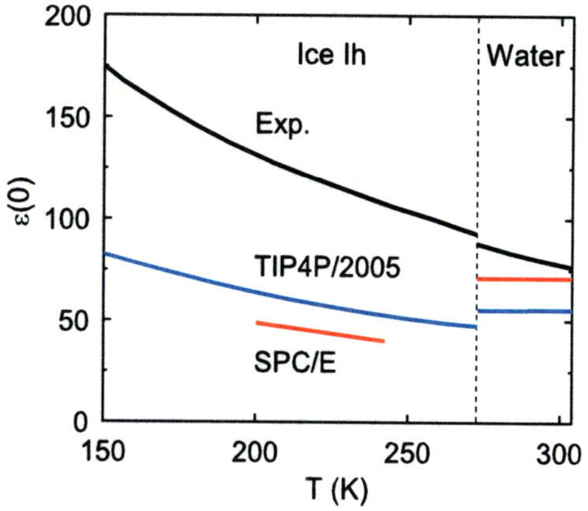

Fig. 1.27 The dielectric constant of ice Ih and water for TIP4P/2005 (blue solid lines) and SPC/E (red solid lines) compared to the experimental values (black solid lines). Data from [112]

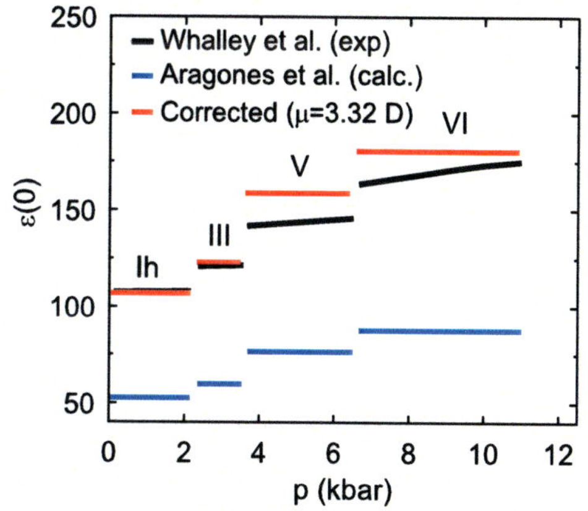

Fig. 1.28 The pressure dependence of the dielectric constant for ices Ih, III, V, and VI at 243 K. The dielectric constant obtained for the TIP4P/2005 model (blue lines) compared to the experimental values (black lines). Corrected values of the dielectric constant for TIP4P/2005 with a dipole moment of 3.32 D (red lines). Data from [112]

(see Fig. 1.28). However, such scaling fails for SPC/E, TIP3P, and TIP5P, as these models predict a different dielectric constant for ice Ih and water at the melting point, thus contradicting the experiment, which shows that they differ by less than 10%.

Tuckerman et al. [113] and Markovich et al. [114] used ab initio methods to model the structure and dynamics of solvated proton complexes which form around H_3O^+ and OH^- ions surrounded by polar water molecules. All atoms, including the excited proton, were considered classical particles. The modeling of RDF (see Sect. 1.2.3) shows that charge transport in water is provided by the Grotthuss mechanism (see Sect. 1.3.1) with an activation energy of 2.4 kcal/mol (0.1 eV), which approximately coincides with the experimental NMR data obtained earlier by Luz and Meiboom [115], as well as with the results of other calculations [116, 117], which yield an activation energy of 0.11–0.14 eV. It is assumed that the proton transport occurs through the interconversion of the Eigen ($H_9O_4^+$) and Zundel ($H_5O_2^+$) cations. Diffusion of the OH^- ion is carried out by analogy through the interconversion of the $H_7O_4^-$ and $H_3O_2^-$ anions. The authors found that the characteristic lifetimes of hydrated ions $H_9O_4^+$ and $H_7O_4^-$ lay in the 2–3 ps time interval, which correspond to that found by means of dielectric spectroscopy [118] (see Chap. 4 for details).

Markovich et al. [119] analyzed the dissociation–association rate of water dimer (H_2O–H_2O) and calculated the self-diffusion coefficient D_w of water molecules. They obtained the value of $D_w = 2.27 \cdot 10^{-9}$ m^2/s, which coincides with the experimental result (see Sect. 1.4). They also found that the dimer recombination–relaxation time is 4.8 ps, which coincides with the second relaxation time of water at room temperature (see Sect. 2.6.2). Similar values for the self-diffusion coefficient were obtained by Liang et al. [117], who also showed that the proton diffusion coefficient is pH-dependent and varies from $D_p = 1.0 \cdot 10^{-9}$ m^2/s at pH = 1 to $D_p = 9.3 \cdot 10^{-9}$ m^2/s at pH = 6. The latter coincides with the proton diffusion coefficient obtained by the equivalent conductivity measurement (see Sect. 1.3.1). The simulated proton exchange rate $k_{pt} = 0.47$ ps^{-1} is close to the experimental value of 0.67 ps^{-1} [120]. It is also approximately equal to the rate $k_d = 0.4$ ps^{-1} that corresponds to the dissociation constant pK_w of water (see Sect. 1.3.2).

There is no standard schematic for the modeling of proton transport and the electrodynamic properties of water and ice. The search for the optimal model is continuing. The limitation of the ab initio approaches is the small number of particles that can be taken into consideration, while DFT requires a priori knowledge of potential. Therefore, a full comparison of the models with the real properties of bulk water is complicated, because a condensed medium is something more than just an ensemble of interacting molecules. Although the qualitative and quantitative description of some properties of water and ice has been achieved [121] and we know how to describe each molecule and the bonds between the pairs of them and how to simulate their motion in a corresponding potential, there is still no real theory of water [122, 123] underpinning a fully predictive model [124], and a universal model of the water molecule has yet to be found [125].

Figure 1.29 illustrates the current status of the search for a universal potential. It shows the ranks of the performance of different water models based on the analysis

Fig. 1.29 An image with the scores from different molecular models for the description of the main electrical, thermodynamic, and structural properties of water. For details, see the text. Reproduced from [92] with permission from the PCCP Owner Societies

proposed by Vega and Abascal [92]. The test of different models showed that the scores (out of a maximum score of 10) range from about 3 (for TIP3P) to about 7 (for TIP4P/2005). None of the water models is perfect. The TIP4P/2005 model gets the maximum score and can be regarded as a variation on the early model of Bernal and Fowler (see Fig. 1.2), which, after 90 years of tests, is still used to model water, although it is becoming obvious that water cannot be "built" using the long-lived molecular species only. We should search for a model which accounts for the intermolecular dynamics in the form of local chemical reactions.

DC conductivity, the dielectric constant, the infrared spectrum, and some geometric parameters of water and ice have been modeled separately. However, these models are highly specialized and very complex. The dynamics of various forms of excess protons, such as H_3O^+, $H_9O_4^+$, or proton holes such as OH^- allows the reproduction of parts of the water dielectric response. For the further developments of computational methods and to account for quantum effects that do not follow from the classical model of Bernal and Fowler, the interconversion of ions and neutral water molecules should be taken into account.

1.7 Summary of the "Structure" of Water and Ice

Analysis has shown that the current structural models follow the original ideas of Kohlrausch [8], Röntgen [7], Bernal and Fowler [9], which are based on static conductometric measurements and early X-ray diffraction data, with two basic assumptions: that water consists of long-lived H_2O molecules with a small concentration of intrinsic ions and that the structure of water is similar to that observed in quartz. All later models, except a few alternative ideas [48], are based on these assumptions.

The standard view on the structure of water can be described as follows (see Fig. 1.30). There is a three-dimensional network of bounded H_2O molecules. The network is based on the tetrahedral coordination of water molecules caused by their polarity and structure. There are defects in the hexagonal structure corresponding to the tetrahedral coordination. Hydrogen atoms do not necessarily occupy the positions according to Bernal–Fowler ice rules. In water, there is significant proton disorder. Molecules are arranged, but they constantly change neighbors in such a way that the average lifetime of the stationary states is about 1 ps. The translational diffusion of molecules occurs via the interstitial mechanism, when a molecule temporally occupies the cavity in the nearest hexagonal ring.[30] A spontaneous fluctuation-born proton transfer between water molecules leads to the formation of ionic species,

[30]In normal conditions, in 1 second, molecules of ice and water were displaced by 0.1 mm and 0.2 μm, respectively. Distances are calculated using the Nernst–Einstein relation and the self-diffusion coefficients $D_W = 2.3 \cdot 10^{-9}$ (298 K) and $D_I = 1.0 \cdot 10^{-14}$ (272 K) m^2/s for water and ice, respectively. These macroscopic distances are much larger than the molecular diameter, and assume continuous molecular shuffling that destroys any long-living molecular clusters.

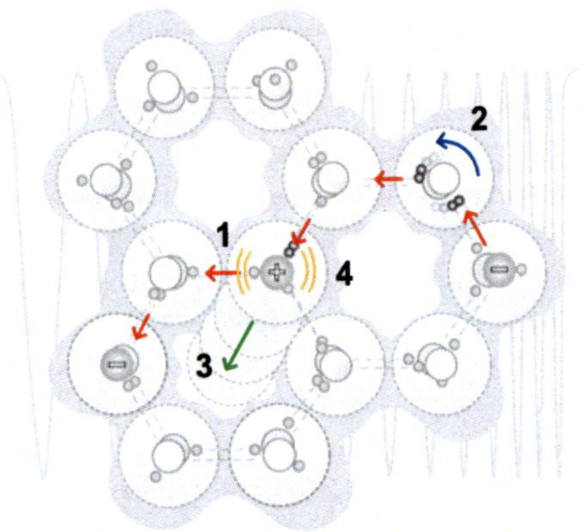

Fig. 1.30 Elements of the dynamic structure of water. Numbers depict: (1) proton transport, (2) molecular adjustment, (3) interstitial diffusion, and (4) molecular/ionic vibrations, respectively

most of which recombine in 0.1 ps, but some of them can live significantly longer. The migration of the excess charge passes through the interconversion of Eigen and Zundel cations.

Despite the fact that the individual properties of water and ice are well described by variations of the Bernal–Fowler model, none of the existing models is capable of explaining a sufficient set of properties necessary for a final description of the dynamical structure. Dielectric relaxation, DC conductivity, the dielectric constant, and infrared absorption are explained separately, but using specific models. Table 1.5 shows several common misconceptions related to the structure of water, which originate from the imperfection of the Bernal–Fowler approach. These misconceptions contradict modern experimental data but are still used in some water models.

Recently, several experimental results were obtained which showed that the Bernal–Fowler model of water fails to reproduce the properties of water, at short (picosecond) time intervals, and also those for water at nanoscales. These results change our current understanding of water's structure and dynamics, and require a significant reconsideration of our vision of water. There are several remarkable examples of the anomalous properties of water, which show that specific details of the water structure and its dynamics are still far from completely understood.

The spatial heterogeneity of water. Small-angle X-ray scattering (SAXS) demonstrates the presence of density fluctuations in ambient water on a physical length scale of about 1 nm [127]. The independent measurements of the time-resolved optical Kerr effect shows characteristic features near 50 and 200 cm^{-1}, which indicate the coexistence of two local structures with high density and low density [16]. This effect is explained in Sect. 4.4 (see Fig. 4.6) within the ionic model of water.

The time heterogeneity of water. A femtosecond pump–probe spectroscopy study of the OH-stretching mode of HDO dissolved in D_2O shows that the orientational relaxation of the HDO molecules occurs in either 13 ps or 0.7 ps, which suggests that two distinct molecular species exist in liquid water with respect to orientational dynamics [128]. A parallel analysis of DC and terahertz conductivity of water reveals two time intervals, which correspond to different interaction potentials, thus exhibit-

Table 1.5 The main misconceptions on the microscopic structure of water and their explanation

Misconception	Experimental data	Comment
The lifetime of an H_2O molecule in liquid water is 11 h	NMR experiments and terahertz spectroscopy show that lifetime of H_2O is less than a microsecond due to fast proton exchange [15, 120]	There is no reason to believe that condensed phases of water consists of long-lived H_2O molecules. It is a dynamic system where protons are not necessarily pinned to oxygen atoms
Hydrogen and hydroxide ions have abnormally high mobilities compared to other ions	The diffusion coefficients of isotopically labeled oxygen and hydrogen atoms are the same in water and ice [63, 64] (see Sect. 1.4)	The proton charge is not necessarily associated with the selected proton. Due to the relay-race mechanism, proton charge translates four times faster than the proton as a mass
Dielectric relaxation, dielectric permittivity, and microwave heating are caused by the orientational motion of polar H_2O molecules	The relaxation time is too long for molecular reorientations; in the time of dielectric relaxation, each molecule passes several molecular distances; if all molecular dipoles were to simultaneously rotate, the expected dielectric constant would be several times lower than the experimental one [126]	Dielectric relaxation is associated with the dynamics of a superlattice of short-lived ionic species [15] (see Sect. 4.5)

ing the temporal heterogeneity of water [15]. The origin of the time heterogeneity of water and ice is discussed in Sect. 3.5. Similar structure features were recently found in alcohols [129].

Anomalies of water in confined geometry. Measurements of water flowing through aligned two-nanometer carbon nanotubes shows that the flow rate exceeds values calculated from continuum hydrodynamic models by more than three orders of magnitude [130]. A dielectric spectroscopy study of water confined in purified nanodiamond ceramic shows that interfacial water has anomalously high protonic conductivity, five orders of magnitude higher than for bulk water [32]. For an explanation of the anomalous properties of water in strong confinement, see Sect. 5.2.

Anomalies of water in an external electric field. A low-frequency electric field drives a water droplet on a solid hydrophobic surface at speed of 10 cm/s caused by the redistribution of electrostatic charge density between the moisture left on the substrate surface and the moving droplet [131]. Other experiments show that pure water forms a free-hanging "bridge" of a few centimeters in the space between two reservoirs under high voltage, revealing properties of water that had been hidden [132] (see Sect. 5.5 for details). The interaction of high-power 3 μm infrared laser radiation with water leads to the spatial separation and motion of electric charges [133], and the emission of a high concentration of positively charged ions [134]. For the explanation and other properties of water in external electric field, see Chap. 5.

Non-molecular species in water. A semi-classical simulation of the molecular dynamics of water shows that water contains charged pairs of concentrations as high as 1 mol/l that participate in the Grotthuss proton-diffusion mechanism [135]. Spectral-weight analysis of the experimental infrared spectra of liquid light (H_2O), heavy (D_2O), and semi-heavy (HDO) water reveals a high (up to 2%) concentration of excess protons (short-lived H_3O^+ ions) [136]. A neutron diffraction experiment with isotopic H/D substitution on a concentrated HCl/H_2O solution shows that the interaction with the solutes affects the tetrahedral network of water molecules and that both dissociated H^+ and Cl^- ions and undissociated HCl molecules coexist in the sample [137]. The latter goes beyond the classical Arrhenius electrolytic dissociation concept, but is confirmed by the parallel analysis of AC and DC electrical conductivities of aqueous electrolyte solutions [54] (see Sect. 5.1).

References

1. J.L.F. Abascal, C. Vega, A general purpose model for the condensed phases of water: TIP4P/2005. J. Chem. Phys. **123** (2005)
2. B.V. Ramirez, R.M. Benito, J. Torres-Arenas, A.L. Benavides, Water phase transitions from the perspective of hydrogen-bond network analysis. Phys. Chem. Chem. Phys. **20**, 28308–28318 (2018)
3. D. Eisenberg, W. Kauzmann, *The Structure and Properties of Water* (Oxford University Press, Oxford, 1969)
4. O. Mishima, L.D. Calvert, E. Whalley, An apparently first-order transition between two amorphous phases of ice induced by pressure. Nature **314**, 76–78 (1985)
5. A.K. Soper, M.A. Ricci, Structures of high-density and low-density water. Phys. Rev. Lett. **84**, 2881–2884 (2000)
6. R. Shi, J. Russo, H. Tanaka, Common microscopic structural origin for water's thermodynamic and dynamic anomalies. J. Chem. Phys. **149**, 224502–21 (2018)
7. W.K. Röntgen, Ueber die Constitution des flüssigen Wassers. Ann. Phys. **45**, 91–97 (1892)
8. F. Kohlrausch, Ad. Heydweiller, Ueber reines Wasser. Ann. d. Phys. u. Chem. **53**, 14–235 (1894)
9. J.D. Bernal, R.H. Fowler, A theory of water and ionic solution, with particular reference to hydrogen and hydroxyl ions. J. Chem. Phys. **1**, 515–548 (1933)
10. R. Mecke, Das Rotationsschwingungsspektrum des Wasserdampfes. I. Zeitschrift für Physik. **81**, 313–331 (1933)
11. P. Atkins, J. de Paula, *Physical Chemistry*, 9th edn. (W. H. Freeman and Company, New York, 2010)
12. K. Amann-Winkel, M.-C. Bellissent-Funel, L.E. Bove, T. Loerting, A. Nilsson, A. Paciaroni, D. Schlesinger, L. Skinner, X-ray and neutron scattering of water. Chem. Rev. **116**, 7570–7589 (2016)
13. P.G. Debenedetti, Supercooled and glassy water. J. Phys. Cond. Matter. **15**, 1669–1726 (2003)
14. A. Nilsson, L.G.M. Pettersson, The structural origin of anomalous properties of liquid water. Nat. Comm. **6**, 2998–3009 (2015)
15. V.G. Artemov, A unified mechanism for ice and water electrical conductivity from direct current to terahertz. Phys. Chem. Chem. Phys. **21**, 8067–8072 (2019)

16. A. Taschin, P. Bartolini, R. Eramo, R. Righini, R. Torre, Evidence of two distinct local structures of water from ambient to supercooled conditions. Nat. Comm. **4**, 2401–8 (2013)
17. P. Perakis, K. Amann-Winkel, F. Lehmkühler, M. Sprung, D. Mariedahl, J.A. Sellberg, H. Pathak, A. Späh, F. Cavalca, D. Schlesinger, A. Ricci, A. Jain, B. Massani, F. Aubree, C.J. Benmore, T. Loerting, G. Grübel, L.G.M. Pettersson, A. Nilsson, Diffusive dynamics during the high-to-low density transition in amorphous ice. PNAS. **114**, 8193–8198 (2017)
18. Ya. Frenkel, *Kinetic Theory of Liquids* (Clarendon Press, Oxford, 1946)
19. F. Rinne, Das kristallsystem und das Achsenverhältnis des Eises. Ges. Wiss. Math. Phys. **69**, 57–62 (1917)
20. W.H. Bragg, The crystal structure of ice. Proc. Phys. Soc. London **34**, 98–103 (1921)
21. D. Starodub, P. Rez, G. Hembree, M. Howells, D. Shapiro, H. Chapman, P. Fromme, K. Schmidt, U. Weierstall, R.B. Doak, J.C.H. Spence, Dose, exposure time and resolution in serial X-ray crystallography. J. Synch. Rad. **15**, 62–73 (2007)
22. M.F. Chaplin, Structure and properties of water in its various states, in *Encyclopedia of Water: Science, Technology, and Society*, ed. P. A. Maurice (Wiley, 2019)
23. P.G. Owston, Diffuse scattering of X-rays by ice. Acta Crystallogr. **2**, 222–228 (1949)
24. A. Kiselev, F. Bachmann, P. Pedevilla, S.J. Cox, A. Michaelides, D. Gerthsen, T. Leisner, Active sites in heterogeneous ice nucleation-the example of K-rich feldspars. Science **355**, 367–371 (2017)
25. P.G. Owston, The structure of ice-I, as determined by x-ray and neutron diffraction analysis. Adv. Phys. **7**, 171–188 (1958)
26. N. Bjerrum, Structure and properties of ice. Science **115**, 385–390 (1952)
27. K. Itagaki, Self-diffusion in single crystal ice. J. Phys. Soc. Jpn. **22**, 427–431 (1967)
28. R. Hanaoka, S. Takata, Y. Nakagami, Electrical treeing in hexagonal ice crystals under applied impulse voltage. IEEE Trans. Diel. Electr. Insul. **11**, 939–945 (2004)
29. L. Pauling, The structure and entropy of ice and of other crystals with some randomness of atomic arrangement. J. Am. Chem. Soc. **57**, 2680–2684 (1935)
30. L. Pauling, *The Nature of Chemical Bond*, 3th ed (Cornell University Press, New York, 1960)
31. H.R. Zelsmann, Temperature dependence of the optical constants for liquid H_2O and D_2O in the far IR region. J. Molecul. Struct. **350**, 95–114 (1995)
32. V.G. Artemov, E. Uykur, P.O. Kapralov, A. Kiselev, K. Stevenson, H. Ouerdane, M. Dressel, Anomalously high proton conduction of interfacial water. J. Phys. Chem. Lett. **11**, 3623–3628 (2020)
33. G.N.I. Clark, C.D. Cappa, J.D. Smith, R.J. Saykally, T. Head-Gordon, Invited topical review: the structure of ambient water. Mol. Phys. **108**, 1415–1433 (2010)
34. M. Leetmaa, K.T. Wikfeldt, M.P. Ljungberg, M. Odelius, J. Swenson, A. Nilsson, L.G. Pettersson, Diffraction and IR/Raman data do not prove tetrahedral water. J. Chem. Phys. **129** (2008)
35. A.N. Tikhonov, V.Y. Arsenin, *Solutions of Ill-Posed Problems* (Winston, New York, 1977)
36. T. Head-Gordon, G. Hura, Water structure from scattering experiments and simulation. Chem. Rev. **102**, 2651–2670 (2002)
37. G. Camisasca, H. Pathak, K.T. Wikfeldt, L.G.M. Pettersson, Radial distribution functions of water: models vs experiments. J. Chem. Phys. **151**, 044502–9 (2019)
38. H. Takeuchi, M. Nakagawa, T. Saito, T. Egawa, K. Tanaka, S. Konaka, X-ray scattering by water molecules studied by using synchrotron radiation. Int. J. Quant. Chem. **52**, 1339–1348 (1994)
39. A. Guinier, *X-ray Diffraction In Crystals, Imperfect Crystals, and Amorphous Bodies* (Dover, New York, 1963)
40. A.K. Soper, Tests of the empirical potential structure refinement method and a new method of application to neutron diffraction data on water. Molecul. Phys. **99**, 1503–1516 (2001)
41. S. Katzoff, X-ray studies of the molecular arrangement in liquids. J. Chem. Phys. **2**, 841 (1934)
42. A.K. Soper, The radial distribution functions of water as derived from radiation total scattering experiments: is there anything we can say for sure? ISRN Phys. Chem. **2013**, 1–67 (2013)

43. P. Wernet, D. Nordlund, U. Bergmann, M. Cavalleri, M. Odelius, H. Ogasawara, L.A. Naslund, T.K. Hirsch, L. Ojamae, P. Glatzel et al., The structure of the first coordination shell in liquid water. Science **304**, 995–999 (2004)
44. F. Sciortino, A. Geiger, H.E. Stanley, Effect of defects on molecular mobility in liquid water. Nature **354**, 218–221 (1991)
45. O.Ya. Samoilov, *Structure of Aqueous Electrolyte Solutions and the Hydration of Ions* (Consultants Bureau, New York, 1965)
46. D.G. Archer, P. Wang, The dielectric constant of water and Debye-Hückel limiting law slopes. J. Phys. Chem. Ref. Dat. **19**, 371–411 (1990)
47. N.W. Ashcroft, N.D. Mermin, *Solid State Physics* (Saunders College, Philadelphia, 1976)
48. V.G. Artemov, Dielectric spectrum of water as a proton dynamics response. Bull. Lebedev Phys. Inst. **42**, 187–191 (2015)
49. A.P. Gaiduk, T.A. Pham, M. Govoni, F. Paesani, G. Galli, Electron affinity of liquid water. Nat. Comm. **9**, 247–6 (2018)
50. D. Marx, Proton transfer 200 years after von Grotthuss: insights from ab initio simulations. Chem. Phys. Chem. **7**, 1848–1870 (2006)
51. J. O'M. Bockris, A.K.N. Reddy, M. Gamboa-Aldeco, *Modern Electrochemistry* (2nd ed.) (Springer, Berlin, 1998)
52. F. Kohlrausch, A. Heydweiller, Ueber reines wasser. Weid. ann. **53**, 209–224 (1894)
53. F. Kohlrausch, A. Heydweiller, Zeit Physik. Chem. **14**, 317–330 (1894)
54. V.G. Artemov, A.A. Volkov, N.N. Sysoev, A.A. Volkov, Conductivity of aqueous HCl, NaOH and NaCl solutions: Is water just a substrate? EPL. **109**, 26002–6 (2016)
55. R. Pauliukaite, J. Juodkazytė, R. Ramanauskas, Theodor von Grotthuss' contribution to electrochemistry. Electrochim. Acta **236**, 28–32 (2017)
56. F. Dahms, R. Costard, E. Pines, B.P. Fingerhut, E.T.J. Nibbering, T. Elsaesser, The hydrated excess proton in the Zundel Cation HO: the role of ultrafast solvent fluctuations. Angew. Chem. Int. Ed. **55**, 10600–10605 (2016)
57. D. Svozil, P. Jungwirth, Cluster model for the ionic product of water: accuracy and limitations of common density functional methods. J. Phys. Chem. A. **110**, 9194–9199 (2006)
58. C. Kobayashi, S. Saito, I. Ohmine, Mechanism of fast proton transfer in ice: potential energy surface and reaction coordinate analyses. J. Chem. Phys. **113**, 9090–9100 (2000)
59. W. Amir, G. Gallot, F. Hache, S. Bratos, J.-C. Leicknam, R. Vuilleumier, Time-resolved observation of the Eigen cation in liquid water. J. Chem. Phys. **126**, 034511–7 (2007)
60. S. Scheiner, Proton transfers in hydrogen-bonded systems. 4. Cationic dimers of ammonia and OH_2. J. Phys. Chem. **86**, 376–382 (1982)
61. N. Agmon, Proton solvation and proton mobility. Israel J. Chem. **39**, 493–502 (1999)
62. E. Pines, D. Huppert, N. Agmon, Geminate recombination in excited-state proton-transfer reactions-numerical-solution of the Debye-Smoluchowski equation with backreaction and comparison with experimental results. J. Chem. Phys. **88**, 5620–5630 (1988)
63. J.H. Wang, C.V. Robinson, I.S. Edelman, Self-diffusion and structure of liquid water: III. Measurement of the self-diffusion of liquid water with H^2, H^3 and O^{18} as tracers. J. Am. Chem. Soc. **75**, 466–470 (1953)
64. K. Goto, T. Hondoh, A. Higashi, Determination of diffusion coefficients of self-interstitials in ice with a new method of observing climb of dislocations by X-ray topography. Jpn. J. Appl. Phys. **25**, 351–357 (1986)
65. M. Holz, S.R. Heil, A. Sacco, Temperature-dependent self-diffusion coefficients of water and six selected molecular liquids for calibration in accurate ^{1}H NMR PFG measurements. Phys. Chem. Chem. Phys. **2**, 4740–4742 (2000)
66. S.A. Fischer, B.I. Dunlap, D. Gunlycke, Correlated dynamics in aqueous proton diffusion. Chem. Sci. **9**, 7126–7132 (2018)
67. M. Francl, Urban legends of chemistry. Nat. Chem. **2**, 600–601 (2010)
68. Quantities and units—Part 8: Physical chemistry and molecular physics, Annex C (normative): pH. International Organization for Standardization (1992)
69. R.G. Bates, *Determination of pH: Theory and Practice* (Wiley, 1973)

70. N.M. Sergeev, N.D. Sergeyeva, Proton exchange in solutions of water in nitromethane and dioxane. Phys. Chem. Chem. Phys. **4**, 2994–2999 (2002)
71. T.R. Dyke, K.M. Mack, J.S. Muenter, The structure of water dimer from molecular beam electric resonance spectroscopy. J. Chem. Phys. **66**, 498–510 (1977)
72. V.G. Artemov, A.A. Volkov, N.N. Sysoev, A. A. Volkov Jr., On autoionization and pH of liquid water. Doklady Phys. **61**, 1–4 (2016)
73. A. Hassanali, M.K. Prakash, H. Eshet, M. Parrinello, On the recombination of hydronium and hydroxide ions in water. PNAS **108**, 20410–5 (2011)
74. K. Itagaki, Self-diffusion in single crystals of ice. J. Phys. Soc. Jpn. **19**, 1081–1081 (1964)
75. J.H. Wang, Self-diffusion and structure of liquid water. I. measurement of self-diffusion of liquid water with deuterium as tracer. J. Am. Chem. Soc. **73**, 510–513 (1951)
76. K. Kreuer, Proton conductivity: materials and applications. Chem. Mater. **8**, 610–641 (1996)
77. P.L. Geissler, C. Dellago, D. Chandler, J. Hutter, M. Parrinello, Autoionization in liquid water. Science **291**, 2121–2124 (2001)
78. K.J. Tielrooij, R.L.A. Timmer, H.J. Bakker, M. Bonn, Structure dynamics of the proton in liquid water probed with terahertz time-domain spectroscopy. Phys. Rev. Lett. **102** (2009)
79. B. Halle, G. Karlstrom, Prototropic charge migration in water. J. Chem. Soc. Faraday Trans. **79**, 1031–1046 (1983)
80. M. Beé, *Quasielastic Neutron Scattering* (Adam Hilger, Bristol, 1988)
81. D.I. Page, J.G. Powles, The correlation of molecular orientation in liquid water by neutron and x-ray scattering. Mol. Phys. **21**, 901–926 (1971)
82. J.G. Powles, J.C. Dore, D.I. Page, Coherent neutron scattering by light water (H_2O) and a light-heavy water mixture (64 per cent H_2O/36 per cent D_2O). Mol. Phys. **24**, 1025–1037 (1972)
83. K.S. Singwi, A. Sjölander, Diffusive motions in water and cold neutron scattering. Phys. Rev. **119**, 863–871 (1960)
84. S.M. Iskanderov, A.G. Novikov, Investigation of the rotational diffusion of water molecules. Russ. J. Phys Chem. T. LVI **10**, 2396–2404 (1982). [in Russian]
85. G.G. Malenkov, Structure and dynamics of liquid water. J. Struct. Chem. **47**, 1–31 (2006)
86. V.S. Oskotsky, On the theory of quasi-elastic scattering of cold neutrons in a liquid. Phys. Condens. Matter. **5**, 1082 (1963). [in Russian]
87. J.A. Barker, R.O. Watts, Structure of water; A Monte Carlo calculation. Chem. Phys. Lett. **3**, 144–145 (1969)
88. A. Rahman, F.H. Stillinger, Molecular dynamics study of liquid water. J. Chem. Phys. **55**, 3336–3359 (1971)
89. A. Wallqvist, R.D. Mountain, Molecular models of water: derivation and description, in *Reviews in Computational Chemistry*, vol. 13, eds by K.B. Lipkowitz, D.B. Boyd (Wiley, Hoboken, NJ, USA, 2007)
90. S. Kale, J. Herzfeld, Natural polarizability and flexibility via explicit valency: the case of water. J. Chem. Phys. **136**, 084109–084118 (2012)
91. W. Kohn, L.J. Sham, Self-consistent equations including exchange and correlation effects. Phys. Rev. A. **140**, 1133–1138 (1965)
92. C. Vega, J.L.F. Abascal, Simulating water with rigid non-polarizable models: a general perspective. Phys. Chem. Chem. Phys. **13**, 19663–19688 (2011)
93. P.Y. Ren, J.W. Ponder, Temperature and pressure dependence of the AMOEBA water model. J. Phys. Chem. B. **108**, 13427–13437 (2004)
94. H.A. Stern, F. Rittner, B.J. Berne, R.A. Friesner, Combined fluctuating charge and polarizable dipole models: application to a five-site water potential function. J. Chem. Phys. **115**, 2237–2251 (2001)
95. P.L. Silvestrelli, M. Parrinello, Water molecule dipole in the gas and in the liquid phase. Phys. Rev. Lett. **82**, 3308–3311 (1999)
96. S.W. Petterson, H.A. Levy, A single-crystal neutron diffraction study of heavy ice. Acta Crystallogr. **10**, 70–76 (1957)

97. S.E. Lappi, B. Smith, S. Franzen, Infrared spectra of $H_2^{16}O$, $H_2^{18}O$ and D_2O in the liquid phase by single-pass attenuated total internal reflection spectroscopy. Spectrochim. Acta, Part A **60**, 2611–2619 (2004)
98. A.A. Kornyshev, Double-layer in ionic liquids: paradigm change? J. Phys. Chem. B. **111**, 5545–5557 (2007)
99. M.J. Gillan, D. Alfé, A. Michaelides, Perspective: how good is DFT for water? J. Chem. Phys. **144** (2016)
100. J.D. Eaves, J.J. Loparo, C.J. Fecko, S.T. Roberts, A. Tokmakoff, P.L. Geissler, Hydrogen bonds in liquid water are broken only fleetingly. Proc. Nat. Sci. Acad. **102**, 13019–13022 (2005)
101. C. Drechsel-Grau, D. Marx, Collective proton transfer in ordinary ice: local environments, temperature dependence and deuteration effects. Phys. Chem. Chem. Phys. **19**, 2623–2635 (2017)
102. R. Car, M. Parrinello, Unified approach for molecular dynamics and density-functional theory. Phys. Rev. Lett. **55**, 2471–2474 (1985)
103. I.F.W. Kuo, C.J. Mundy, M.J. McGrath, J.I. Siepmann, J. VandeVondele, M. Sprik, J. Hutter, B. Chen, M.L. Klein, F. Mohamed, M. Krack, M. Parrinello, Liquid water from first principles: investigation of different sampling approaches. J. Phys. Chem. B. **108**, 12990–12998 (2004)
104. K. Laasonen, M. Sprik, M. Parrinello, R. Car, "Ab initio" liquid water. J. Chem. Phys. **99**, 9080–9089 (1993)
105. M.E. Tuckerman, D. Marx, M.L. Klein, M. Parrinello, On the quantum nature of the shared proton in hydrogen bonds. Science **275**, 817–820 (1997)
106. M.E. Tuckerman, D. Marx, M. Parrinello, The nature and transport mechanism of hydrated hydroxide ions in aqueous solution. Nature **417**, 925–929 (2002)
107. D. Asthagiri, L.R. Pratt, J.D. Kress, M.A. Gomez, Hydration and mobility of HO^-_{aq}. PNAS **101**, 7229–7233 (2004)
108. S. Yoo, X.C. Zeng, S.S. Xantheas, On the phase diagram of water with density functional theory potentials: the melting temperature of ice IhIh with the Perdew-Burke-Ernzerhof and Becke-Lee-Yang-Parr functionals. J. Chem. Phys. **130–4**, 2211020 (2009)
109. W.C. Natzle, C.B. Moore, Recombination of hydrogen ion (H+) and hydroxide in pure liquid water. J. Phys. Chem. **89**, 2605–2612 (1985)
110. M. Eigen, L. De Maeyer, Untersuchungen über die Kinetik der Neutralisation. Z. Elektrochem. **59**, 986–993 (1955)
111. S.W. Rick, A.D.J. Haymet, Dielectric constant and proton order and disorder in ice Ih: Monte Carlo computer simulations. J. Chem. Phys. **118**, 9291–9296 (2003)
112. J.L. Aragones, L.G. MacDowell, C. Vega, Dielectric constant of ices and water: a lesson about water interactions. J. Phys. Chem. A. **115**, 5745–5758 (2011)
113. M. Tuckerman, K. Laasonen, M. Sprik, M. Parrinello, Ab initio molecular dynamics simulation of the solvation and transport of hydronium and hydroxyl ions in water. J. Chem. Phys. **103**, 150–161 (1995)
114. O. Markovitch, H. Chen, S. Izvekov, F. Paesani, G.A. Voth, N. Agmon, Special pair dance and partner selection: elementary steps in proton transport in liquid water. J. Phys. Chem. B. **112**, 9456–9466 (2008)
115. Z. Luz, S. Meiboom, The activation energies of proton transfer reactions in water. J. Am. Chem. Soc. **86**, 4768–4769 (1964)
116. A.A. Kornyshev, A.M. Kuznetsov, E. Spohr, J. Ulstrup, Kinetics of proton transport in water. J. Phys. Chem. B. **107**, 3351–3366 (2003)
117. C. Liang, T.L.C. Jansen, Proton transport in a binary biomimetic solution revealed by molecular dynamics simulation. J. Chem. Phys. **135**, 114502–8 (2011)
118. A. A. Volkov, V.G. Artemov, A.A. Volkov Jr., N.N. Sysoev, Possible mechanism of molecular motion in liquid water from dielectric spectroscopy data. J. Mol. Liq. **248**, 564–568 (2017)
119. O. Markovitch, N. Agmon, Reversible geminate recombination of hydrogen-bonded water molecule pair. J. Chem. Phys. **129**, 084505–13 (2008)

120. S. Meiboom, Nuclear magnetic resonance study of the proton transfer in water. J. Chem. Phys. **34**, 375–388 (1961)
121. C.H. Cho, S. Singh, G.W. Robinson, Understanding all of water's anomalies with a nonlocal potential. J. Chem. Phys. **107**, 7979–7988 (1997)
122. B. Cabane, R. Vuilleumier, The physics of liquid water. Comptes Rendus Geosci. Elsevier **337**, 159 (2005)
123. G.R. Ivanitskii, A.A. Deev, E.P. Khizhnyak, Long-term dynamic structural memory in water: can it exist? Phys.-Usp. **57**, 37–65 (2014)
124. V.P. Sokhan, A.P. Jones, F.S. Cipcigan, J. Crain, G.J. Martyna, Signature properties of water: their molecular electronic origins. PNAS. **112**, 6341–6346 (2015)
125. B. Guillot, A reappraisal of what we have learnt during three decades of computer simulations of water. J. Mol. Liq. **101**, 219–260 (2002)
126. A. von Hippel, The dielectric relaxation spectra of water, ice, and aqueous solutions, and their interpret at ion 1. Critical Surv. Status-quo Water, IEEE Tnans. Electric. Insul. **23**, 801–816 (1988)
127. C. Huang, K.T. Wikfeldt, T. Tokushima, D. Nordlund, Y. Harada, U. Bergmann, M. Niebuhr, T.M. Weiss, Y. Horikawa, M. Leetmaa, M.P. Ljungberg, O. Takahashi, A. Lenz, L. Ojamäe, A.P. Lyubartsev, S. Shin, L.G.M. Pettersson, A. Nilsson, The inhomogeneous structure of water at ambient conditions. PNAS **106**, 15214–15218 (2009)
128. S. Woutersen, U. Emmerichs, H.J. Bakker, Femtosecond Mid-IR Pump-Probe spectroscopy of liquid water: evidence for a two-component structure. Science **278**, 658–660 (1997)
129. V.G. Artemov, A. Ryzhov, E. Carlsen, P. Kapralov, H. Ouerdane, Nonrotational mechanism of polarization in alcohols. J. Phys. Chem. B. **124**, 11022–11029 (2020)
130. J.K. Holt, H.G. Park, Y. Wang, M. Stadermann, A.B. Artyukhin, C.P. Grigoropoulos, A. Noy, O. Bakajin, Fast mass transport through Sub-2-nanometer carbon nanotubes. Science **312**, 1034–1037 (2006)
131. M. Gunji, M. Washizu, Self-propulsion of a water droplet in an electric field. J. Phys. D Appl. Phys. **38**, 2417–2423 (2005)
132. E.C. Fuchs, Can a century old experiment reveal hidden properties of water? Water **2**, 381–410 (2010)
133. N.N. Il'ichev, L.A. Kulevskii, P.P. Pashinin, Spatial separation and motion of electric charges arising due to the interaction of high-power IR laser radiation with water. Quant. Electron. **43**, 47–54 (2013)
134. D.V. Klochkov, G.E. Kotkovskii, A.S. Nalobin, E.S. Tananina, A.A. Chistyakov, Ion formation upon water excitation by IR laser radiation in the range of OH stretching vibrations. JETP Lett. **75**, 20–22 (2002)
135. C. Bai, J. Herzfeld, Special pairs are decisive in the autoionization and recombination of water. J. Phys. Chem. B. **121**, 4213–4219 (2017)
136. V.G. Artemov, E. Uykur, S. Roh, A.V. Pronin, H. Ouerdane, M. Dressel, Revealing excess protons in the infrared spectrum of liquid water. Sci. Rep. **10**, 11320–9 (2020)
137. A. Botti, F. Bruni, S. Imberti, M.A. Ricci, A.K. Soper, Ions in water: the microscopic structure of a concentrated HCl solution. J. Chem. Phys. **121**, 7840–7848 (2004)

Chapter 2
The Interaction of Electromagnetic Waves with Water

Abstract Water is the most important substance in our everyday life, and has been studied as no other medium. While it has a relatively simple atomic composition, water nevertheless presents an astonishing variety of manifestations of its interaction with electromagnetic waves of the different wavelengths from radio frequencies to X-rays, representing its uniqueness compared to other dielectrics. In this chapter, the response of water to external electromagnetic radiation is considered in an extended frequency range from 0 to 10^{15} Hz. The independent view on the separate parts of the spectrum, such as the dielectric constant, DC conductivity, radio wave, microwave, terahertz, and IR absorption, is provided along with a comprehensive view on the entire spectrum as a whole by means of sum rules, Kramers–Kronig analysis and the isotope effect. A curious reader will find a complete description of water's electrodynamic parameters, as well as specific questions regarding the dynamic structure of water.

2.1 Maxwell's Equations in the Presence of Water

On the macroscopic level, water is a dielectric which can be considered an ensemble of charged particles (electrons and atomic nuclei) which interact with each other and with an external electric field. The interaction of water with electromagnetic waves has the same nature as that for classic dielectrics, and can be described by Maxwell–Lorentz equations, which in the most general case have the following form:

$$\begin{cases} rot\,\mathbf{h} = \frac{4\pi}{c}\rho\mathbf{v} + \frac{1}{c}\frac{\partial \mathbf{e}}{\partial t}, \\ rot\,\mathbf{e} = -\frac{1}{c}\frac{\partial \mathbf{h}}{\partial t}, \\ div\,\mathbf{h} = 0, \\ div\,\mathbf{e} = 4\pi\rho, \end{cases} \tag{2.1}$$

where ρ and $\mathbf{v}$ are the charge density and velocity, c is the speed of light, and $\mathbf{e}$ and $\mathbf{h}$ are microscopic rapidly changing electric and magnetic fields, respectively. All quantities are space and time dependent.

V. Artemov, *The Electrodynamics of Water and Ice*, Springer Series in Chemical Physics 124, https://doi.org/10.1007/978-3-030-72424-5_2

The system of equations, given by (2.1), can be completed by the system of equations of motion applied for electrons and atom nuclei:

$$m\mathbf{a} = q\mathbf{e}\,(\mathbf{v}, t) + \frac{q}{c}\,[\mathbf{v}, \mathbf{h}\,(r, t)] + \mathbf{f}, \tag{2.2}$$

where m is mass, q is electric charge, and $\mathbf{f}$ is the force.

Although systems (2.1) and (2.2) are complete and allow one to describe any atomic-molecular ensemble, they cannot be applied to the real macroscopic substance, because the microscopic fields and charge distribution change in time (in the range of 10^{-10}–10^{-15} s) and in space (at distances of about 0.1–1.0 Å). Thus, a detailed microscopic analysis would require high computational costs. For practical use, Maxwell suggested averaging, and replacing all quantities by the averaged quantities, thus replacing system (2.1) by the following:

$$\begin{cases} rot\,\mathbf{B} = \frac{4\pi}{c}\rho\mathbf{v} + \frac{1}{c}\frac{\partial\mathbf{E}}{\partial t}, \\ rot\,\mathbf{E} = -\frac{1}{c}\frac{\partial\mathbf{B}}{\partial t}, \\ div\mathbf{B} = 0, \\ div\mathbf{E} = 4\pi\ \langle\rho\rangle, \end{cases} \tag{2.3}$$

where $\mathbf{E} = \langle e\rangle$ and $\mathbf{B} = \langle h\rangle$ are the averaged vectors of the electric and magnetic fields, respectively. Although this system of equation loses some information about the local fields, it requires much lower computational costs and can be applied to any continuous medium.

Within this approach, the magnitude of the electric induction $\mathbf{D}$ is connected to the dielectric polarization tensor $\mathbf{P}$ and the external electric field $\mathbf{E}$ by the following formula:

$$\mathbf{D} = \mathbf{E} + 4\pi\mathbf{P}, \tag{2.4}$$

or in a more general time-dependent form:

$$\mathbf{D}(t) = \mathbf{E}(t) + \int_0^\infty f(\tau)\mathbf{E}(t-\tau)d\tau, \tag{2.5}$$

where $f(\tau)$ is a dielectric response function which characterizes the properties of the media and τ represents the delay of the reaction of the medium in a changing external electric field. Inasmuch as any alternating field can be reduced to a set of monochromatic waves by the Fourier transform, the time dependencies of all quantities have a factor of $e^{-i\omega t}$, where $\omega = 2\pi\nu$ and ν are angular and normal frequencies, respectively. For such fields, the relation between $\mathbf{D}$ and $\mathbf{E}$ is

$$\mathbf{D} = \varepsilon(\omega)\mathbf{E}, \tag{2.6}$$

where the frequency-dependent function $\epsilon(\omega)$ is the dielectric function of the medium, which is a function of its thermodynamic state and obeys the dispersion law:

$$\varepsilon(\omega) = 1 + \int_0^\infty f(\tau) e^{i\omega\tau} d\tau. \tag{2.7}$$

The function $\epsilon(\omega)$ is a complex quantity, which assumes that it can be presented in the following form:

$$\varepsilon(\omega) = \varepsilon'(\omega) + i \cdot \varepsilon''(\omega) = \varepsilon'(\omega) + i\frac{4\pi\sigma'}{\omega}, \tag{2.8}$$

where $i = \sqrt{1}$ is the unit imaginary number, ϵ' and ϵ'' are the real and imaginary parts of the dielectric function, respectively, and σ' is the dynamic conductivity.[1]

Following the causality principle, the real and imaginary parts of the dielectric constant satisfy the integral Kramers–Kronig relations:

$$\varepsilon'(\omega) - 1 = \frac{1}{\pi} \int_{-\infty}^{+\infty} \frac{\varepsilon''(x)}{x - \omega} dx = \frac{2}{\pi} \int_0^{+\infty} \frac{x\varepsilon''(x)}{x^2 - \omega^2} dx, \tag{2.10}$$

$$\varepsilon''(\omega) = -\frac{1}{\pi} \int_{-\infty}^{+\infty} \frac{\varepsilon'(x) - 1}{x - \omega} dx. \tag{2.11}$$

Assuming in (2.10) that $x \ll \omega$ and comparing result with (2.9), we get the so-called sum rule [1]:

$$\frac{m}{2\pi^2 e^2} \int_0^\infty \omega\varepsilon''(\omega) d\omega = \int_0^\infty f(\omega) d\omega = N, \tag{2.12}$$

where $f(\omega)d\omega$ is the oscillator strength.

---

[1]Note that in classical electrodynamics it is customary to divide currents into the conduction and displacement components, writing the dielectric function as follows:

$$\varepsilon(\omega) = i\frac{\sigma_{dc}}{\varepsilon_0\omega} + F(\omega), \text{ (2.9)}$$

where σ_{dc} is the direct current conductivity and $F(\omega)$ is a frequency-dependent dielectric function. However, (2.9) is incorrect and contains a wrong physical meaning. Since both the conduction current and the displacement currents are indistinguishable for the infinite sample and obviously have the same nature, hereinafter, by $\sigma = \epsilon''\epsilon_0\omega$, where ϵ_0 being the vacuum permittivity, we mean the frequency-dependent dynamic conductivity, which includes both DC and AC conductivity parts, and assume that formula (2.5) is a definition of the electrical conductivity function, which is valid at any frequency (including $\omega = 0$).

Hence, the integral (or area under the curve of the frequency-dependent dynamic conductivity $\sigma(\omega)$) is proportional to the concentration of the charges per unit, and one can find this concentration has a broadband spectrum (see Sect. 2.7 for further details).

For dielectrics at $\omega \to 0$, formula (2.10) is transformed into

$$\varepsilon'(0) - 1 = \frac{2}{\pi} \int_0^{\infty} \frac{\varepsilon''(x)}{x} dx, \tag{2.13}$$

from which one obtains an expression for the static dielectric constant:

$$\varepsilon(0) - 1 = \frac{4\pi e^2}{m} \int_0^{\infty} \frac{f(\omega)}{\omega^2} d\omega. \tag{2.14}$$

Thus, the static dielectric constant $\epsilon(0)$ of a dielectric material is defined by the frequency-dependent dielectric response (see Sect. 2.4 for water).

In the linear-response approximation, and assuming that electromagnetic waves in the medium are plane and monochromatic, the square of the wave vector is equal to

$$\mathbf{k}^2 \equiv k'^2 - k''^2 + 2i\mathbf{k}'\mathbf{k}'' = \varepsilon\mu\frac{\omega^2}{c^2}, \tag{2.15}$$

from which for a medium with absorption, we have

$$k = \sqrt{\varepsilon\mu}\omega/c, \tag{2.16}$$

where

$$\sqrt{\varepsilon\mu} = n + i\kappa, \tag{2.17}$$

and n is the refractive index of the medium, κ is the imaginary part of the refractive index, which determines the rate of wave decay. The coefficient κ is related to the absorption α = 1/x · lg (I_0/I), where I_0 and I are the intensities of the incident and transmitted-through-a-layer-x radiation, respectively, by

$$\alpha = \frac{4\pi\kappa}{\lambda}, \tag{2.18}$$

where λ is the wavelength in the medium.

For water $\mu = 1$, we obtain the relation between n and κ with the real and imaginary parts of the dielectric constant equal to

$$\varepsilon' = n^2 - \kappa^2, \ \varepsilon'' = 2n\kappa. \tag{2.19}$$

Equations (2.16)–(2.19) connect all the main terms of the spectroscopic data, and allow a recalculation of the corresponding frequency-dependent quantities into each other, as according to local traditions, the different parts of the water spectrum is represented in different terms, which complicates its comprehensive analysis. Hereinafter, we use terms of the complex dielectric function $\epsilon(\omega) = \epsilon'(\omega) + i\epsilon''(\omega)$ and dynamic conductivity, $\sigma = \epsilon''\epsilon_0\omega$, as the most convenient form of representation of the broadband dielectric response of water and ice. Note that $\epsilon(\omega)$ directly contributes to (2.6), and thus represents the simplest form of the data representation, regardless of the technique used for its determination.

2.2 The Broadband Dielectric Spectroscopy of Water

Although $\epsilon(\omega)$ fully characterizes the material and is a good parameter for analysis, it cannot be measured directly. One should first understand the basics of dielectric spectroscopy, in order to see how the dielectric response is measured in different frequency ranges.

Figure 2.1 shows various methods used to determine the electrodynamic parameters of water and ice, which can be recalculated as the real ϵ' and the imaginary ϵ'' parts of the dielectric function. In the low-frequency range (0–10^7 Hz), the complex impedance $Z^*(\omega)$ of the sample[2] is measured in a parallel-plate capacitor circuit. In the linear response approximation, assuming that the sample is homogeneous and isotropic, and that the field changes by the harmonic law, the dielectric function $\epsilon(\omega)$ can be found by the following formula:

$$\varepsilon(\omega) = 1/(i\omega Z^*(\omega)C_0), \tag{2.20}$$

where C_0 is the capacitance of the vacuum.

In the middle-frequency range (10^6–10^{11} Hz), the function $\epsilon(\omega)$ is derived by measuring the complex propagation coefficient (reflection or transmission). In this case, a waveguide or cavity circuits are used, and the impedance of the sample is calculated by the following formula:

$$Z^*(\omega) = Z_0\frac{1 + r^*(l)}{1 - r^*(l)}, \tag{2.21}$$

where l is the length of the measuring line, $r^*(l)=U^*_{ref}/U^*_{inc}$ is a complex reflection factor, where U^*_{ref} and U^*_{inc} are the reflected and the incident signals, respectively.

[2]Note that the sample is not a material, because it always has finite size and boundaries, which can affect the dielectric measurement at low frequencies (see [2]). Thus, one should be careful with the transfer of the sample parameters to the material properties, and use the corresponding equivalent schemes.

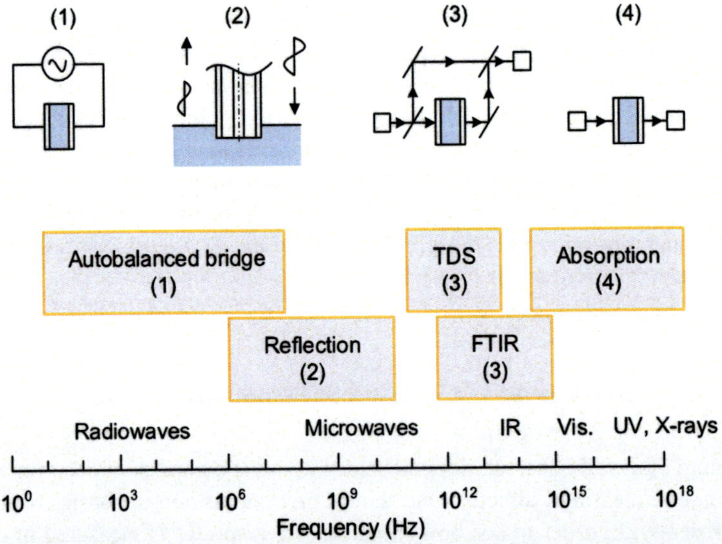

Fig. 2.1 The range of methods for the analysis of the dielectric response of water and ice from radio waves to X-rays, which includes impedance analysis (10^1–10^7 Hz); the network wave analysis (10^7–10^{11} Hz); time-domain terahertz spectroscopy (TDS) (10^{11}–10^{12} Hz); IR-Fourier spectroscopy (10^{13}–10^{14} Hz); optical spectroscopy (about 10^{15} Hz); and ultraviolet and X-ray spectroscopy (10^{15}–10^{18} Hz). Insets show the main measurement schematics: (1) parallel-plate capacitor; (2) open-end reflection; (3) Mach–Zehnder interferometry; (4) quasi-optical absorption measurement. The water/ice sample is shown in blue

In the terahertz and IR frequency ranges (10^{10}–10^{14} Hz), the dielectric parameters are measured by a Mach–Zehnder interferometer or a cavity resonator. The latter is used more for low-temperature measurements. The dielectric function is calculated from the measured transmittance and phase shift using Fresnel formulas (see, for example, [3]). A significant problem for optical measurements is the exclusion of cuvette windows for water, and the boundary layers for ice. The attenuated total reflection (ATR) method avoids this problem, but also has disadvantages related to anomalous dispersion and the non-absolute-value problem.

The measurement accuracy differs for different parts of the spectrum and depends on the size and shape of the sample, the quality of the electrodes, and the measurement technique. Modern devices allows one to achieve the following accuracy, which is given in terms of the dielectric loss tangent, δ [4][3]: tg $\delta < 10^{-3}$ (0–10^7 Hz); tg $\delta < 10^{-2}$ (10^6–10^{11} Hz); tg $\delta < 10^{-2}$ (10^{10}–10^{15} Hz). However, if the measurement method allows one to measure one of the parameters, $\epsilon'(\omega)$ or $\epsilon''(\omega)$, with better quality, the accuracy can be increased by using Kramers–Kronig relations (2.10), and (2.11).

[3]The loss tangent is defined as the ratio (or angle in a complex plane) of the lossy reaction to the electric field E in the curl equation to the lossless reaction: tg $\delta = \epsilon''/\epsilon'$.

The listed modes of non-perturbing (an approximation of a linear response) sensing of water cover the range from intramolecular vibrations in the IR region, with characteristic times of the order of tens of femtoseconds, to structural relaxations that take hours, and are combined under the general name of ultra-broadband dielectric spectroscopy. The broadband dielectric spectroscopy method has an advantage over other methods (for example, inelastic neutron scattering, nuclear magnetic resonance, and isotope substitution (see Chap. 1)), because it allows one to observe a wide time window of the dielectric polarization effects. All types of structures of water from instantaneous (I-structure) to diffusion averaged (D-structure) (see Sect. 1.1) are available for analysis by this technique. The main limitation of this method is the integrity of measurements and the difficulty of the unambiguous identification of contributions from various mechanisms, such as molecular reorientations, charge separation, and the flickering or drift of the charge. This limitation, however, is tempered by analysis of the temperature, phase transition, and isotope effects on the broadband spectrum.

Figure 2.2 shows the generalized ultra-broadband dielectric response of water, collated from experimental data, using (2.15)–(2.19). The spectrum covers 14 orders of the frequency magnitude, and includes the radio wave, microwave, terahertz, infrared, optical, and ultraviolet regions. In terms of the dielectric function (upper panel), the spectrum is characterized by the main feature: the Debye relaxation, and by the static dielectric constant $\epsilon(0)$. The dynamic conductivity spectrum, $\sigma(\omega)$, consists of two parts which partially overlap in the visible region: the low-frequency atomic contribution (blue) and high-frequency electronic (yellow). The low-frequency limit corresponds to DC conductivity (see Sect. 1.3). Particular details of the spectrum shown in Fig. 2.2 are discussed in Sects. 2.3–2.8 for water and Sects. 3.2–3.5 for ice. The microscopic interpretation of the spectrum is given in Chap. 4. More spectral data on ice and water can be found in appendix.

2.3 Microwave Spectrum: Dielectric Relaxation

2.3.1 Experimental Data

The dielectric (Debye) relaxation band is the main part of the broadband dielectric spectrum of water shown in Fig. 2.2, and one of the most fundamental properties that characterize its molecular dynamics. The phenomenon of dielectric relaxation in water has been extensively studied for several decades, and is based on the individual properties of the dielectric function and the specific structural properties of water (see Sect. 4.5.1).

In terms of permittivity $\epsilon^*(\omega) = \epsilon'(\omega)+i\epsilon''(\omega)$ (see Fig. 2.2 upper panel), water shows a wide bell-shaped band (see $\epsilon''(\omega)$) and the corresponding changes in $\epsilon'(\omega)$. In other words, water shows frequency variation in the dielectric parameters or dispersion. If the external field frequency is low, water simply reduces the strength of

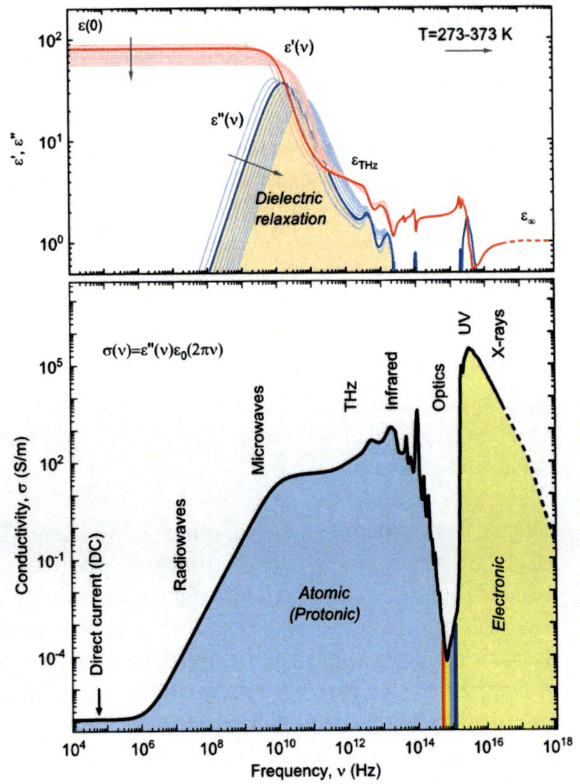

Fig. 2.2 The ultra-broadband dielectric spectra of water: real, ϵ', and imaginary, ϵ'', parts of the dielectric function (upper panel) and dynamic conductivity, $\sigma = \epsilon''\epsilon_0\omega$ (lower panel) in the temperature interval from 273 to 373 K (0–100 °C). Colored areas correspond to: dielectric (Debye) relaxation (orange), atomic (protonic) contribution (blue), and electronic contribution (yellow). Bold curves are for 20 °C

the electric field, by a factor of 80, which is the static dielectric constant, $\epsilon(0)$ (see the next section). Here, the dielectric losses $\epsilon''(0)$ are negligibly small and correspond to the static DC conductivity (see the bottom panel in Fig. 2.1). However, when the electric field is changing relatively fast, the dielectric constant $\epsilon'(\omega)$ reduces (see red curve in the upper panel), as the molecular system becomes unable to follow the changes of the electric field. A delay in polarization P in the response of water to the changing electric field E appears. The time lag between E and P implies the absorption of electromagnetic radiation caused by an irreversible degradation of free energy and leads to the attenuation of electromagnetic waves passing through the medium, showing, in this way, the relaxation band shown in Fig. 2.3 (see blue curve in the upper panel). This band was named after Peter Debye, who made the first detailed description of this phenomena [5].

Debye found that the microwave dielectric spectrum of water, which includes the dielectric relaxation band (orange area in Fig. 2.2), is described by the following formula:

$$\varepsilon(\omega) = \varepsilon'(\omega) + i\varepsilon''(\omega) = \varepsilon_\infty + \frac{\Delta\varepsilon}{1 + i\omega\tau_D}, \tag{2.22}$$

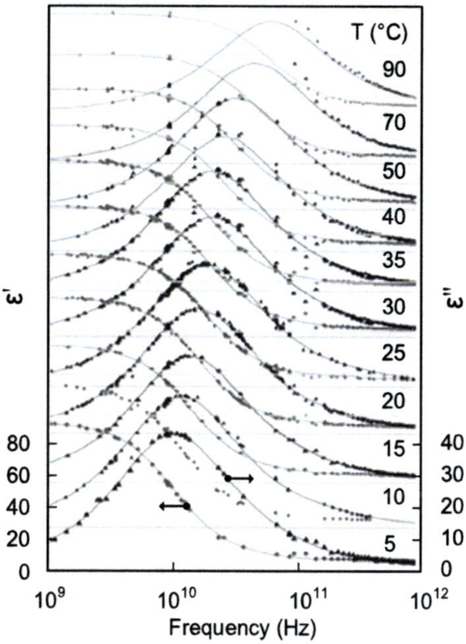

Fig. 2.3 The microwave part of the real, ϵ', and imaginary, ϵ'', parts of the dielectric function of water at different temperatures from 5 to 90 °C (see numbers near the curves). Graphs for temperatures above 5 °C are shifted up for clarity. Dots are experimental data points, lines are model (2.25). The dielectric relaxation band shows a blue shift with an increase in temperature. Collated on the data from [6]

where ϵ_∞ is the high-frequency limit of the dielectric constant, τ_D is the relaxation time, and $\Delta\epsilon = \epsilon(0) - \epsilon_\infty$, where $\epsilon(0)$ is the static dielectric constant of water.[4] Dividing the real and imaginary parts, one gets

$$\varepsilon'(\omega) = \varepsilon_\infty + \frac{\Delta\varepsilon}{1 + \omega^2\tau_D^2}; \quad (2.23)$$

$$\varepsilon''(\omega) = \frac{\Delta\varepsilon \cdot \omega\tau_D}{1 + \omega^2\tau_D^2}. \quad (2.24)$$

Formula (2.22) fits the experimental data for water up to 100 GHz, but at higher frequencies, the experimental dielectric loses exceed those predicted by the Debye equation, demonstrating an additional contribution. This high-frequency wing of Debye relaxation is discussed in Sect. 2.6.

Figure 2.3 shows the temperature dependence of the dielectric relaxation mode. As the temperature increases, the relaxation band shifts in the direction of higher frequencies (blue shift). The band intensity and the corresponding dielectric constant, $\epsilon(0)$, decrease with an increase in temperature, showing the negative temperature trend of the dielectric constant (see Fig. 2.12).

---

[4] Equation (2.22) has an important consequence. It connects the dielectric relaxation band with the static dielectric constant, which means that they both have the same microscopic nature discussed in Sect. 4.5.

Generalized dielectric spectra of water allow for several decomposition options. The most simple way to treat the dielectric response is to use an additive sum of a Lorentz-type oscillatory and the Debye-type relaxation functions. Following this approach, Ellison [6] analyzed more than 3,000 experimental data points and derived analytical formulas which fit the experimental water spectrum in an extended frequency range from 1 MHz to 25 THz, and a temperature interval from 273 to 373 K. He used three relaxors and two oscillators, according to the following formulas:

$$\varepsilon'(\omega, T) = \varepsilon(0, T) - \sum_{n=1}^{3} \frac{\omega^2 \tau_n^2(T) \Delta\varepsilon_n(T)}{1 + \omega^2 \tau_n^2(T)}$$
$$- \sum_{n=4}^{5} \frac{\omega^2 \tau_n^2(T) \Delta\varepsilon_n(T)[1 + \tau_n^2(T)(\omega^2 + \omega_0^2)]}{[1 + \tau_n^2(T)(\omega_0 + \omega)^2][1 + \tau_n^2(T)(\omega_0 - \omega)^2]}; \qquad (2.25)$$

$$\sigma(\omega, T) = \sigma_{dc}(T) + \sum_{n=1}^{3} \frac{\Delta\sigma_n(T) \cdot \omega^2 \tau_n^2(T)}{1 + \omega^2 \tau_n^2(T)}$$
$$+ \sum_{n=4}^{5} \frac{\Delta\sigma_n(T) \cdot \omega^2 \tau_n^2(T)}{2 \cdot [1 + \tau_n^2(T)(\omega_n + \omega)^2][1 + \tau_n^2(T)(\omega_n - \omega)^2]}, \qquad (2.26)$$

where

$$\varepsilon(0, T) = 87.9144 - 0.404399(T - 273) + 9.58726 \cdot 10^{-4}(T - 273)^2 - 1.32802 \cdot 10^{-6}(T - 273)^3, \qquad (2.27)$$

$$\ln(\sigma_{dc}^{-1}) = \alpha_0 + \alpha_1 T + \alpha_2 T^2 + \alpha_3 T^3 + \alpha_4 T^4 + \alpha_5 T^5, \qquad (2.28)$$

and $\alpha_0 = 4,45656$, $\alpha_1 = -7,3309 \cdot 10^{-2}$, $\alpha_2 = 5,0273 \cdot 10^{-4}$, $\alpha_3 = -2,5792 \cdot 10^{-6}$, $\alpha_4 = 6,6206 \cdot 10^{-9}$ and $\alpha_5 = 7,0484 \cdot 10^{-13}$. For n=1,2,3:

$$\Delta\sigma_n(T) = \varepsilon_0 \Delta\varepsilon_n(T)/\tau_n(T), \qquad (2.29)$$
$$\Delta\varepsilon_n(T) = a_n \exp(-b_n(T - 273)), \qquad (2.30)$$
$$\tau_n(T) = c_n \exp(d_n/(T + 406.2883 - 2 \cdot 273)), \qquad (2.31)$$

and for n = 4 and 5:

$$\Delta\varepsilon_4(T) = p_0 + p_1(T-273) + p_2(T-273)^2, \quad (2.32)$$

$$\omega_4(T)/2\pi = p_3 + p_4(T-273) + p_5(T-273)^2 + p_6(T-273)^3, \quad (2.33)$$

$$\tau_4(T) = p_7 + p_8(T-273) + p_9(T-273)^2 + p_{10}(T-273)^3, \quad (2.34)$$

$$\Delta\varepsilon_5(T) = p_{11} + p_{12}(T-273) + p_{13}(T-273)^2, \quad (2.35)$$

$$\omega_5(T)/2\pi = p_{14} + p_{15}(T-273) + p_{16}(T-273)^2, \quad (2.36)$$

$$\tau_5(T) = p_{17} + p_{18}(T-273) + p_{19}(T-273)^2. \quad (2.37)$$

The coefficients a_i, b_i, c_i, d_i, where $i = 1-3$, and the coefficients T_c and p_n, where n = 0–19, are given in Table 2.1. The characteristic parameters of the spectrum at different temperatures, calculated by formulas (2.25) and (2.26), are given in Table 2.2.

Figure 2.4 shows the structure of the microwave spectrum of water decomposed according to formulas (2.25)–(2.37). The function exhaustively describes the experi-

Table 2.1 Coefficients for the formulas (2.31)–(2.37)

$a_1 = 79.2388$	$b_1 = 4.3006 \cdot 10^{-4}$	$c_1 = 1.3823 \cdot 10^{-13}$	$d_1 = 652.7648$
$a_2 = 3.8159$	$b_2 = 1.1174 \cdot 10^{-2}$	$c_2 = 3.5104 \cdot 10^{-16}$	$d_2 = 1249.533$
$a_3 = 1.6350$	$b_3 = 6.8415 \cdot 10^{-3}$	$c_3 = 6.3004 \cdot 10^{-15}$	$d_3 = 405.5169$
$p_0 = 0.8380$	$p_5 = 2.7382 \cdot 10^8$	$p_{10} = 1.655 \cdot 10^{-19}$	$p_{15} = -7.4414 \cdot 10^{10}$
$p_1 = -6.1186 \cdot 10^{-3}$	$p_6 = -1.2469 \cdot 10^6$	$p_{11} = 0.6166$	$p_{16} = 4.9745 \cdot 10^8$
$p_2 = -1.2937 \cdot 10^{-5}$	$p_7 = 9.6186 \cdot 10^{-14}$	$p_{12} = 7.2385 \cdot 10^{-4}$	$p_{17} = 2.8825 \cdot 10^{-14}$
$p_3 = 4.2360 \cdot 10^{12}$	$p_8 = 1.7958 \cdot 10^{-16}$	$p_{13} = -9.5234 \cdot 10^{-6}$	$p_{18} = -3.1421 \cdot 10^{-16}$
$p_4 = -1.4261 \cdot 10^{10}$	$p_9 = -9.3100 \cdot 10^{-18}$	$p_{14} = 1.5983 \cdot 10^{13}$	$p_{19} = 3.5281 \cdot 10^{-18}$

Table 2.2 The parameters of the microwave dielectric spectrum of water in the temperature range from 0 to 100 °C: σ_{dc} and $\epsilon(0)$ being the static (DC conductivity) and the static dielectric constant, respectively; ω_{D1} and τ_{D1} are the frequency and the time of Debye relaxation, respectively; σ_{D1} is the high-frequency conductivity plateau in the GHz region

T (°C)	σ_{dc} (μ S/cm)	$\epsilon(0)$	$\omega_{D1} = 2\pi\nu_{D1}$ (GHz)	$\tau_{D1} = 2\pi/\omega_{D1}$ ($\cdot 10^{-10}$ s)	σ_{D1} (S/cm)
0	0.0118	88.3	8.79	7.14	0.45
10	0.0229	83.9	12.2	5.15	0.64
20	0.0421	80.1	16.4	3.83	0.8
30	0.0710	76.7	21.2	2.96	0.96
40	0.113	72.8	27.1	2.32	1.18
50	0.172	69.9	32.9	1.91	1.36
60	0.249	66.9	39.7	1.58	1.5
70	0.345	63.9	46.8	1.34	1.62
80	0.468	61	54.6	1.15	1.76
90	0.615	58.4	62.5	1.00	1.94
100	0.784	55.7	71.5	0.878	2.17

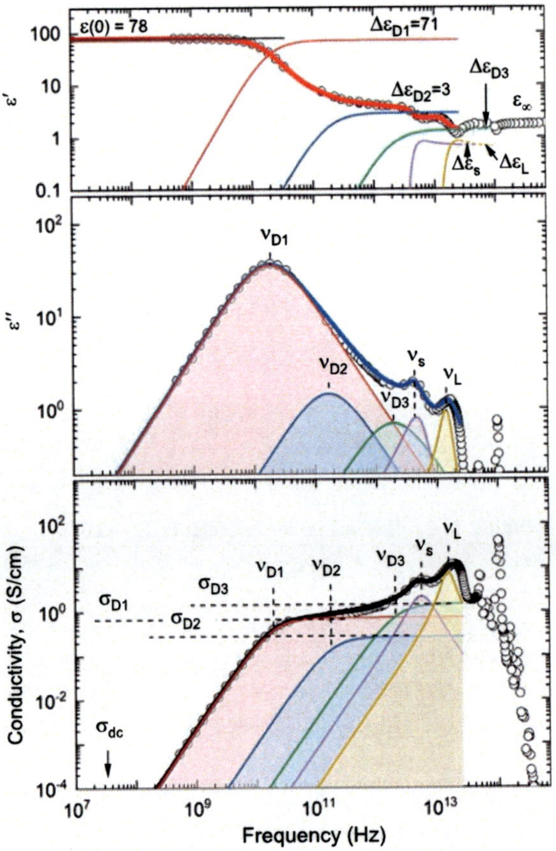

Fig. 2.4 The structure of the microwave spectrum of water at room temperature. Bold lines are the model according to (2.25), (2.26); thin lines are the five additive components of the model. Circles represent the experimental data

mental data points up to 25 THz. The Cole–Cole plot of the dielectric loss, ϵ'', versus the real part of the dielectric permittivity, ϵ', in the whole temperature interval is given in appendix.

A similar fit of the dielectric spectrum of water was obtained by Yada et al., who also got a satisfactory description of both light and heavy water spectra with only two relaxors and two oscillators [7]. Parameters of Yada et al. roughly coincide with those obtained by Ellison. Figure 2.4 shows that the first (main) relaxation band gives 95% of the contribution to the static dielectric constant, $\epsilon(0)$, while the cumulative contribution of all the other components is only a few percent. From the point of view of dynamic conductivity, the relaxation bands are saturated at high frequencies, which contradicts the sum rule (see Sect. 2.8 for details). Moreover, neither Tanaka's nor Ellison's fits account for static DC conductivity.

An alternative fitting function, which is free of these problems, was derived in [8]. The model is based on a single fitting function, and assumes a common molecular mechanism for high and low frequencies (see Sect. 3.5 for details), and additionally accounts for static DC conductivity, which is missing in two previous models.

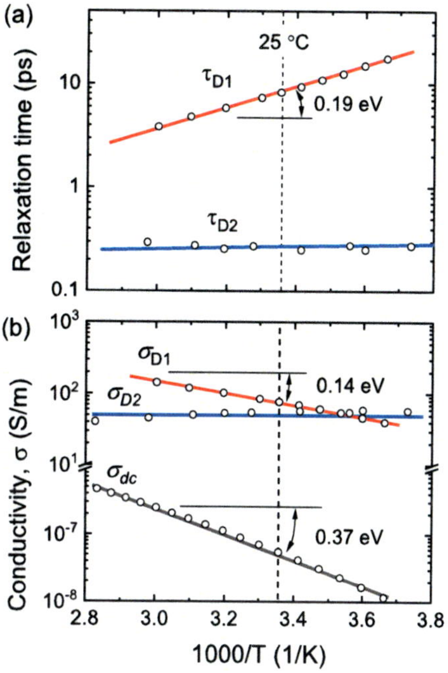

Fig. 2.5 The temperature dependencies of: **a** relaxation times for first, τ_{D1}, and second, τ_{D2}, relaxation bands, and **b** static, σ_{dc}, first high-frequency, σ_{D1}, and second high-frequency, σ_{D2}, conductivities (see Fig. 2.4) for liquid water. Lines are fit with parameters given in Table 2.3. Numbers near curves are activation energies in eV

Apart from different interpretations of the microscopic mechanism of polarization, all three approximations have similar data-fitting accuracy, and can be used for the analytical description of the microwave spectrum of water in different technological applications.

Figure 2.5 shows the temperature dependencies of the main parameters of dielectric relaxation. Two relaxation times: the first, τ_{D1}, and the second, τ_{D2}, correspond to the frequencies of the relaxation-peak maxima, ν_{D1} and ν_{D2}, respectively (Fig. 2.5a). In terms of dynamic conductivity, the relaxation time loses its functionality, being replaced by high-frequency conductivity plateaus, the first, σ_{D1}, and the second, σ_{D2}, are shown in Fig. 2.5b. The plateaus can be defined by the Debye formula as

$$\sigma_{D1} = \frac{\varepsilon_0 \Delta \varepsilon_{D1}}{\tau_{D1}}, \tag{2.38}$$

where ε_0 is the permittivity of vacuum.

The parameters of the first relaxation band, shown in Fig. 2.5, have typical Arrhenius behavior with an activation energy ΔE of both high-frequency conductivity, σ_{D1}, and the relaxation frequency, ν_{D1}, being close to that for the self-diffusion coefficient (see Table 2.3). Thus, the Debye relaxation in water seems to be a diffusion-controlled process.

The second relaxation band parameters, on the contrary, do not depend on temperature. Yada et al. suggested [9] that the high-frequency relaxation time corresponds

Table 2.3 The values of the activation energies ΔE and pre-exponential factors A for the temperature dependencies of the parameters shown in Fig. 2.5 and for the self-diffusion coefficient D_{self}

	σ_{D1} (S/m)	$\tau_{D1}=1/\nu_{D1}$ (ps)	D_{self} (m^2/s)	σ_{dc} (S/m)
A	$1.6{\cdot}10^4$	67	$2.3{\cdot}10^{-6}$	7.90
ΔE (eV)	0.14± 0.03	0.18± 0.03	0.18± 0.02	0.37± 0.05

to the molecular collision time, $\tau_{col} = 1/4{\cdot}\rho\ r_{eff}^2\ \mathrm{sqrt}(\pi^2 m/k_B T)$, where r_{eff} and m are the effective diameter and mass of water molecules, respectively, and ρ is the density of water. Using $\rho = 1000\,\mathrm{kg/m^3}$, $r_{eff} = 1.5$ Å, and $m = 3{\cdot}10^{-26}$ kg, one gets a value which is close to the experimental one. Thus, the second relaxation is a collision-controlled process.

2.3.2 *Data Interpretation*

Great efforts have been made to study the dielectric relaxation of water [6, 10, 11]; however, its microscopic mechanism remains not fully understood, and the existing interpretations are controversial [11–14] (see Sect. 4.5.1 for an alternative microscopic description). Equation (2.22) means the exponential relaxation of polarization can result from very different polarization mechanisms. In other words, Debye relaxation can be obtained from a very general model of the dynamics in a double-well potential.

Figure 2.6 shows the typical model of the double-well potential. Let us assume that the system, following the external electric field, occupies only one of two possible states. We mark the quantities related to these states with indices 1 and 2. The rate at which the concentration of the molecules of the first state changes, due to their transition to the second state, is

$$-\frac{dn_1}{dt} = n_1 k_1 - n_2 k_2, \tag{2.39}$$

where n_1 and n_2 are the populations of the corresponding states and k_1 and k_2 are the specific rates (probabilities) of the forward ($1 \rightarrow 2$) and reverse ($2 \rightarrow 1$) transitions between the states.

If one assumes that f is an external force, which drives the transfer between states, and that this force is weak enough to satisfy the approximation of the linear response,[5] the new populations can be defined from the undisturbed populations without an electric field by the following formulas:

[5]In a linear response, a weak perturbation generates a small out-of-equilibrium response that is proportional to this perturbation. The response is expected to be proportional to this perturbation, where the response coefficient is independent of the strength of the external electric field [15].

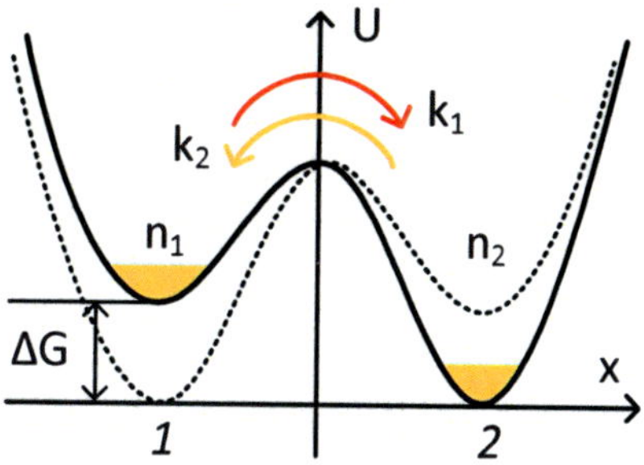

Fig. 2.6 The model of the double-well potential for the dielectric relaxation in water (see the text for description)

$$\begin{cases} n_1 = n_1^0 + f\frac{dn_1}{df},\ k_1 = k_1^0 + f\frac{dk_1}{df}, \\ n_2 = n_2^0 - f\frac{dn_2}{df},\ k_2 = k_2^0 + f\frac{dk_2}{df}. \end{cases} \tag{2.40}$$

In dynamic equilibrium, we get

$$\frac{n_2^0}{n_1^0} = \frac{k_1^0}{k_2^0} = \exp(\Delta G/kT), \tag{2.41}$$

where ΔG is the difference in free energies between the states 1 and 2.

Equation (2.41) gives

$$\frac{dk_1}{df} = k_2^0 \frac{n_2^0}{n_1^0}\left(\frac{1}{kT}\frac{d\Delta G}{df}\right) + \frac{n_2^0}{n_1^0}\frac{dk_2}{df}. \tag{2.42}$$

Substituting (2.40) into (2.42) and assuming that $f = f_0\exp(i\omega t)$, we get

$$\frac{d}{dt}\left(f\frac{dn_1}{df}\right) + (k_1^0 + k_2^0)\left(f\frac{dn_1}{df}\right) = -k_2^0 n_2^0\left(\frac{1}{kT}\frac{d\Delta G}{df}\right) f_0 \exp(i\omega t). \tag{2.43}$$

Equation (2.43) has a solution:

$$-\frac{dn_1}{df} = \frac{n\left(\frac{1}{kT}\frac{dG}{df}\right)}{2\left[1 + ch\left(\frac{\Delta G}{kT}\right)\right]} \frac{1}{1 + \frac{i\omega}{k_1^0 + k_2^0}}, \tag{2.44}$$

which contains all the properties of Debye relaxation. Thus, the Debye form of relaxation is obtained for any mechanism that satisfies the transition in the two-minimum potential with the relaxation time $\tau_D = (k_1^0 + k_2^0)^{-1}$ in the linear-response approximation.

The simple form of the dielectric relaxation with a single relaxation time, τ_D, prompted Debye to conclude that relaxation is determined by the orientations of identical particles. He suggested that dipoles of H_2O can synchronously reorient, following the external electric field, and that the dielectric response depends on time, because the orientation lags with increasing frequency due to viscous friction. Although this simple idea is still popular, it has been shown (see, for example, [11]) that the orientation of H_2O dipoles gives incorrect values of the static dielectric constant of water and fails to reproduce its temperature dependence. Nowadays, it is obvious that the process of dielectric relaxation has more complex collective dynamics [13, 14], which are only partially associated with molecular reorientations (see Sect. 4.5).

Jaccard [16] assumed that dielectric relaxation in ice, which has been shown to be very similar to that in water [17], can be explained by two types of charged defects. First are conventional H_3O^+ and OH^- ions responsible for DC conduction, and second are Bjerrum valence defects, which have either two protons (D defect) or no protons (L defect) between two neighboring molecules, which manifest at high frequencies. This model allows one to qualitatively describe the dielectric spectrum of ice [18], although Bjerrum defects have never been observed experimentally.

Recently, Jaccard's theory has been applied to water, and the main dielectric relaxation has been explained by the dynamics of Bjerrum defects [19]. Although the dielectric spectrum of water has been qualitatively described, the relatively high value of the high-frequency conductivity of water, σ_{D1}, assumes that the defect concentration is as high as a few percent of all water molecules. This means that the strong interaction between *all* charge carriers, which is missing in the proposed model, should be accounted for. However, the authors do not imply any interaction between different types of defects. The interaction should, first, affect the DC conductivity that the authors assumed is due to the *free* motion of H_3O^+ and OH^- ions, which is not the case, and second, give additional relaxation time, which is not observed. Moreover, the lifetime of Bjerrum defects is expected to be about a few picoseconds, as it is determined by molecular reorientations, while the Debye relaxation time is dozens of picoseconds, thus significantly exceeding the expected lifetime of the charge carriers. In other words, the model of two types of charged defects does not fully satisfy the experimental results.

An alternative mechanism for the dielectric relaxation of water was first suggested in [17]. The authors explained static conductivity and Debye relaxation by the Brownian dynamics of interacting H_3O^+ and OH^- ions, thus considering only one type of charge carrier. The high-frequency dielectric response, which includes Debye relaxation, is shown to be the result of the polarization of interacting counterions, while low-frequency DC conductivity is produced by the same ionic species, which are long-lived enough to overcome mutual screening and contribute to static conductivity. Although the concentration of all H_3O^+ and OH^- ions was shown to be high, the effective concentration of those ions participating in DC conductivity is exactly the same as assumed by pH. It was also shown that this model (the ionic model of water presented in Sect. 4.2.3) describes the experimental conductivity spectrum of water and ice from DC current to THz [8]. For a detailed analysis of the ionic model, see Chap. 4.

Thus, the microscopic description of dielectric relaxation in water is still debated. We can say for sure that it does not reduce to the orientations of H_2O dipoles. The rotational mechanism of polarization is fully confirmed for the gas phase (water vapor) only. In the condensed state, however, the model of orientational polarization leads to a lower dielectric constant than observed experimentally (see Sect. 2.4). In addition, it has been shown [20] that the relaxation time, calculated by the Stokes–Einstein formula ($\tau_{D1} = 4\pi r^3 \eta / k_B T$, where r is the molecular radius and η is viscosity) is too large if the macroscopic experimental viscosity is used.[6] The intermolecular polarization mechanism, due to the dynamics of intrinsic ionic species, looks more appropriate for the experimental data. Recently, a similar, nonrotational mechanism of polarization has been discussed in monohydric alcohols, the dielectric properties of which are very similar to those of water [21]. Both water and alcohols are two typical representatives of polar liquids. Just like water, alcohols have a relatively high dielectric constant and are good solvents. As alcohols have longer molecules than those in water, rotational polarization is not possible. Thus, the nonrotational mechanism of polarization suggested for alcohols is also possible for water, as their spectra collapse to a single curve being scaled (see [21]).

2.3.3 Is Debye Relaxation a Unique Feature of Water?

Polarization ability governs various mechanical, chemical, and thermodynamic properties of dielectrics. Detectors, sensors, memory units, and electric-power storage systems all use this property. That is why different polar dielectrics have been studied in detail. Although Debye relaxation is known as a hallmark of water, which is actually missing in most dielectrics, a very similar relaxation band was also detected in a narrow class of materials such as organic solvents, molten salts, superionic conductors, and ionic liquids. In the context of the microscopic dynamics, which accompany dielectric relaxation, it is worth analyzing the spectroscopic data of these materials and compare them with those for water.

Figure 2.7 compares dielectric spectra of water, a molecular liquid (glycerol), an ionic liquid (ethylammonium nitrate), and amorphous quartz (SiO_2). The general similarity of the spectra is observed for all substances except quartz. One can see that the spectrum of SiO_2, which served as an analog of water structure in X-ray diffraction experiments (see Sect. 1.2), does not exhibit Debye relaxation, shown as the yellow-shaded area in Fig. 2.7. Thus, Bernal–Fowler analogy between water and quartz is not complete.

Nevertheless, the Debye relaxation is observed for some dielectrics, such as ferroelectrics, superionic conductors, alcohols, and ionic liquids. All these substances are characterized by the ability of the atomic-molecular structure to rearrange. One can see in Fig. 2.7 that the spectra of glycerol, ethylammonium nitrate, and water are very similar in the microwave frequency range. The analogies between water and

[6] The difference between experimental viscosity and that calculated by Stokes–Einstein formula is observed for all polar liquids and, depending on the conditions, ranges from ten to several thousand times.

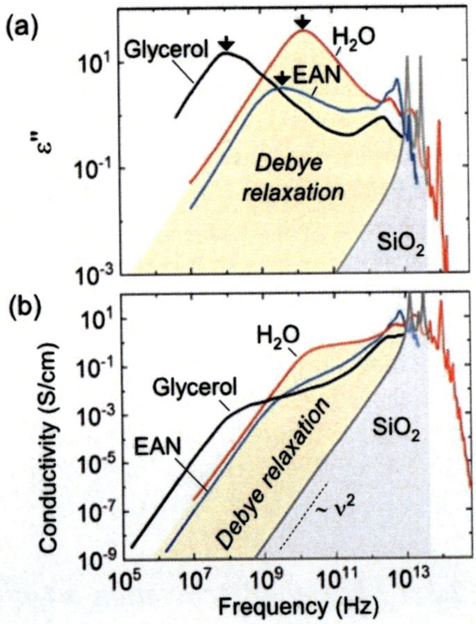

Fig. 2.7 Spectra of **a** the imaginary part, ϵ'', of the dielectric constant (dielectric loses), and **b** the dynamic conductivity, σ, of water (H_2O), glycerol ($C_3H_8O_3$), ethylammonium nitrate ($C_2H_8N_2O_3$), and quartz (SiO_2). Experimental data from [22, 23] and the personal archive of the author are used. The yellow area shows the Debye relaxation region, which is not observed in the spectra of normal dielectrics, such as quartz (compare the gray area). Black arrows show the maxima of the dielectric relaxation for different substances

the other liquids are also manifested in the close activation energy of the relaxation time and similar physical properties, yet the basic mechanisms underpinning their dielectric properties were not fully analyzed in comparison.

2.4 The Static Dielectric Constant

2.4.1 Experimental Data

One of the most recognizable electrodynamic parameters of water is the static dielectric constant, $\epsilon(0)$, which at 25 °C equals 78.4 and increases to 93.0 at –1 °C. This value is relatively high for dielectrics, but yet not the highest (see Table 2.4). For example, such ferroelectric materials as $BaTiO_3$ or $SrTiO_3$ have dielectric constants of 4,000 and 300, respectively, while paraelectric SiO_2, TiO_2, and Teflon have dielectric constants of 4, 50, and 2, respectively. Despite the commonly accepted opinion, the dielectric constant is not always connected with the dipole moment of the molecule. For example, the KCl molecule has dipole moment 6 D[7] (more than three times higher than for the water molecule), while the dielectric constant of KCl crystal is only about 5.

The $\epsilon(0)$ of water is important for various practical applications. It is used in electronics, energy applications, communications, environmental sciences, analytical chemistry, and electrochemistry. This parameter determines the characteristics of the signal-transmission lines, indicates the phase transition and temperature change,

[7] 1 D (Debye) = $3.33 \cdot 10^{-30}$ C·m.

Table 2.4 Static dielectric constant, $\epsilon(0)$, of some materials

Substance	$\epsilon(0)$
Potassium tantalite niobate (KTN)	6000
Barium titanate ($BaTiO_3$)	4000
Potassium niobate ($KNbO_3$)	700
Strontium titanate ($SrTiO_3$)	300
Water	80
Titanium dioxide (TiO_2)	50
Acetone (C_3H_6O)	20
Silicon (Si)	12
GaAs	11
Soda–lime glass	7
KCl	5
Fused silica (SiO_2)	4
Epoxy	4
PVC	3.5
Teflon	2
Paraffin	2
Air	1.0006

determines the solvation ability, and is used as a dielectric reference standard. Most importantly, however, is that it allows us to understand the molecular dynamics of water, because the causality principle and Kramers–Kronig relations (2.10), (2.11) associate $\epsilon(0)$ with dielectric losses, which is represented by the imaginary part of the dielectric function ($\epsilon''(\omega)$), the main part of which is the Debye relaxation associated with molecular dynamics. The Debye relaxation makes up to 95% of the static dielectric constant of water.[8] Thus, the problems of the dielectric relaxation interpretation (see Sect. 2.3) are automatically transferred to the dielectric constant, and vice versa.

Figure 2.8 shows the experimental temperature dependencies of the static dielectric constant of water and ice. Malmberg et al. [24] and Johari et al. [25] showed that these curves are described by the formulas $\epsilon_w(0) = 87.740\text{–}0.40008 \cdot t^\circ + 9.398 \cdot 10^{-4} \cdot t^{\circ 2}\text{–}1.410 \cdot 10^{-6} \cdot t^{\circ 3}$ for water and $\epsilon_I(0) = 3.2 + (24620)/(t^\circ + 266.8)$ for ice, where t° is temperature in °C. It can be seen from the figure that the difference between the values of $\epsilon(0)$ near 273 K for water and ice is less than 10%, and that except of this slight shift, which is apparently due to the density difference (see the solid lines and the right-hand scale), the curves generally show a uniform trend.[9]

[8] The same is true for alcohols (see, for instance, [21]).

[9] This is despite the fact that the relaxation time changes by seven orders of magnitude (see Chap. 4 for details).

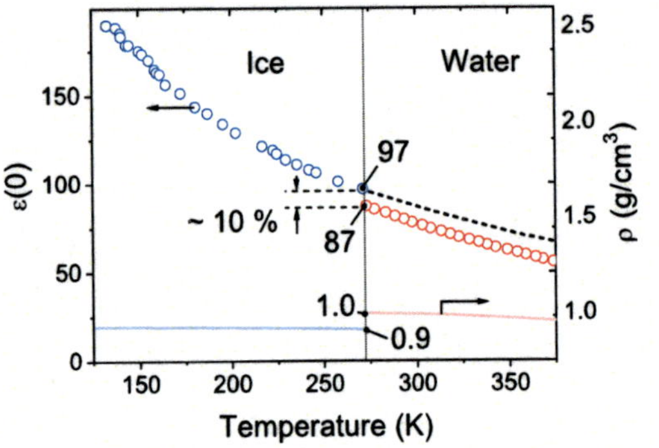

Fig. 2.8 The temperature dependencies of the experimental static dielectric constant, $\epsilon(0)$, of water [24] and ice [25], measured by the AC-bridge technique and the three-terminal coaxial capacitor method, respectively. The right-hand scale is for the density function (solid lines)

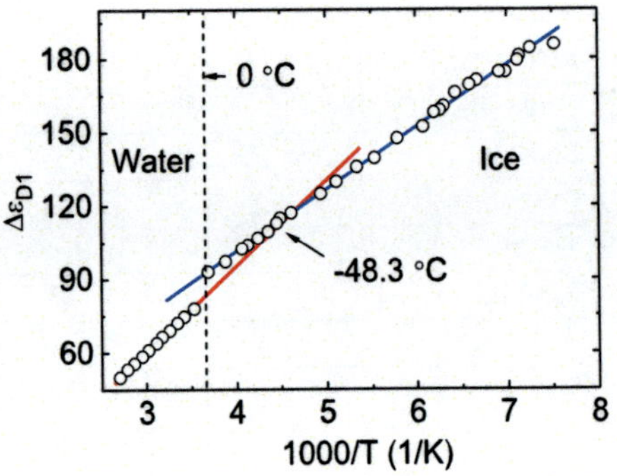

Fig. 2.9 The Debye relaxation contribution $\Delta\epsilon_{D1} = \epsilon(0) - \epsilon_{THz}$ to the static dielectric constant as a function of temperature. Lines are fit according to (2.45)

Figure 2.9 shows the temperature dependencies of the Debye relaxation contributions $\Delta\epsilon_{D1}$ to the total static dielectric constant $\epsilon(0)$. Both curves obey the Curie–Weiss law:

$$\Delta\varepsilon_{D1}(T) = A/T + B, \tag{2.45}$$

where A equals 33,400 and 25,260 K^{-1}, and B equals −40 and 0.20 for water and ice, respectively. The curves intersect each other at the maximum supercooling temperature of water at atmospheric pressure, which is −48.3 °C (225 K) [26]. By coincidence, this temperature corresponds to the maximum of supercooling: below this point water has never been observed in a liquid state at normal pressure.

Figure 2.10 shows the temperature dependencies of the static dielectric constant at different pressures. Interestingly, in the supercritical region, where water is more like a dense gas of H_2O molecules [27], than a liquid, the dielectric constant of water drops down to about 5, which is more than an order of magnitude lower than that under normal conditions. This shows the important role of the cooperativity of water molecules in the dielectric constant, and that the dipole moment of molecular does not play a significant role.

The microscopic interpretation of the static dielectric constant of water has a long history which is not yet complete. It is believed that water and ice have a high

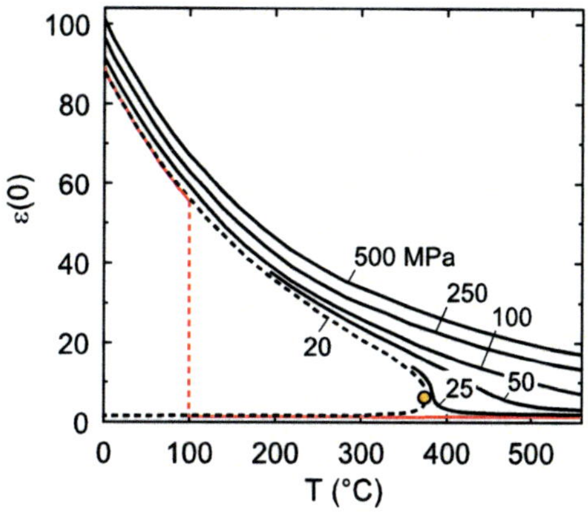

Fig. 2.10 The static dielectric constant of water as a function of temperature at various pressures. The red curve is for ambient pressure. The orange circle shows the critical point. Data from [28]

dielectric constant, since they are capable of rapidly transmitting local fluctuations of polarization [10]. The microscopic nature of such fluctuations is still discussed. There are two basic approaches: the molecular (the local-field approach) and the intermolecular (the spatial-charge-separation approach) which have been applied for the explanation of the dielectric constant of water.

2.4.2 The Local-Field Approach

The local-field approach is based on the notion of the average local field, E', which acts in the interior of a molecule of a dielectric placed in an external electric field E [29]. An approximation of the local field, which has been widely used, is the Lorentz field:

$$E' = E + 4\pi P/3, \tag{2.46}$$

where P is the averaged polarization. Although the Lorentz field has been shown to work for non-polar substances [30], it is entirely inadequate in the case of polar substances, such as water.

Let us consider the development of the model for the dielectric constant of water in the local-field approach and its final collapse. The starting point is the macroscopic Clausius–Mossotti equation, which generally works for any isotropic dielectric [30]:

$$\frac{\varepsilon - 1}{\varepsilon + 2} = \frac{\alpha_M}{3\epsilon_0 V}, \tag{2.47}$$

where α_M=$n\alpha$ is the macroscopic polarizability of a volume$V$ and n is the concentration of unit cells each with a polarizability α.

The first approach follows Debye's idea of the molecular dipole reorientation mechanism [5, 30, 31]. The starting assumption here is that the dielectric constant arises from the ordering of polar H_2O molecules in an external field E, which acts against thermal disordering. Assuming that water molecules are rigid dipoles, μ_0, and that the applied E and the locally acting E' fields are equivalent, one obtains

$$\varepsilon(0) - \varepsilon_\infty = \frac{n\mu_0^2}{3\epsilon_0 k_B T}, \tag{2.48}$$

which is Debye's equation for the static dielectric constant [5]. Equation (2.48) leads to a static dielectric constant which is much smaller than the experimental value. For example, at μ_0=1.85 D and room temperature it gives $\epsilon(0) = 13$.

A more successful calculation of the local field in polar liquids was made by Onsager [32]. He treated the molecule as a real cavity in a statistical continuum of the uniform dielectric constant ϵ equal to that of the liquid in bulk. On the basis of this model, the electrostatic theory of continuous media leads at once to a simple expression for the local field and the average torque effective in orienting a dipole molecule relative to the field E. The molecular dipole moment in Onsager's approach contains a permanent part and a part induced by the local field E', and derived the following formula:

$$\frac{(\varepsilon(0) - \varepsilon_\infty)(2\varepsilon(0) + \varepsilon_\infty)}{\varepsilon(0)(\varepsilon_\infty + 2)^2} = \frac{n\mu_0^2}{9\varepsilon_0 k_B T}, \tag{2.49}$$

which at $\epsilon(0) \gg \epsilon_\infty$ gives

$$\varepsilon(0) - \varepsilon_\infty = \frac{n\mu_0^2}{3\varepsilon_0 k_B T}\frac{(\varepsilon_\infty + 2)^2}{6}. \tag{2.50}$$

For room temperature, μ_0=1.85 D, and $\epsilon_\infty \approx 2$, (2.50) gives the value $\epsilon(0) \approx 30$, which is higher than those obtained by Eq. 2.49, but still several times lower than the experimental one.

Kirkwood [33] suggested an extension of Onsager's theory and assumed that Bernal–Fowler tetrahedrally coordinated molecule (see Sect. 1.2) orients in the dielectric continuum together with its first coordination shell. The field of the center dipole μ_0 causes a favored average orientation of neighboring dipoles and thus an increased moment $\mu = g \cdot \mu_0$ of the tetrahedral group of five H_2O molecules. If we assume a modified Bernal–Fowler structure for water and introduce a g-factor of about 2.7, the experimental value, $\epsilon(0) \approx 78$, of the dielectric constant of liquid water at 25 °C is obtained from (2.50) with a dipole moment in the liquid 26% greater than that in the vapor.

Although the theory described above reduces the problem relating to the dielectric polarization of polar liquids to the problem of the calculation of the g-factor and the sum of the dipole moments of overpolarized molecules in its immediate environment in water, there are still several points which do not allow one to fully accept this theory:

- The concept that a group of five molecules can orient as a unit while retaining an internal freedom of rotation seems incompatible with the notion of an extended tetrahedral network structure [11].
- Using (2.50), the dielectric constant does not reduce to ϵ_∞ at frequencies sufficiently high for the contribution of the permanent dipole, μ_0, orientation to become zero, as expected for scalar isotropic polarizability [10].
- The Debye–Onsager–Kirkwood–Fröhlich theory of the dielectric polarization assumes a dramatic stretching of the water molecule in the liquid state in comparison with the gas phase, which is not confirmed by IR and neutron scattering measurements. For example, an adequate description of the experimental curve is possible (see Fig. 1.27) with the dipole moment μ = 3.32 D, which is 180% larger than the gas-phase dipole moment μ_0 = 1.85 D. However, the comparison of the data from the neutron diffraction of ice [34] and the IR spectroscopy of water vapor [35] shows that the O–H distance changes from 0.099 and 0.096 nm, respectively, showing a less than 4% difference. Furthermore, the IR molecular stretching modes shift from gas to liquid only a few percent toward lower frequencies (see Sect. 2.5).
- There is no uniform model for ice and water. The experimental temperature dependence of the static dielectric constant is achieved with different structural parameters for ice and water [36], while they are very similar from the IR spectroscopy point of view (see Sect. 2.5).
- The static dielectric constant is 95% formed by the dielectric relaxation band (see Sect. 2.3). The self-diffusion coefficients of water and ice are so high (see Sect. 1.4) that during the relaxation time each particle is displaced by several molecular distances, and thus they are oriented and translated many times. In other words, any molecular configurations are destroyed by thermal shuffling over the period of the external field. Diffusion should lead to the averaging of all intermolecular dipole moments.
- In a system of identical dipole moments, individual dipoles tend to lose their identity and "dissolve" in a system of compensating polar charges. The compensation occurs by a change in the electrostatic free energy of the system; therefore, polar liquids with a high Onsager's local-field ordering are expected to be poor solvents, which is not the case for water.

Thus, the Debye–Onsager–Kirkwod–Fröhloch's mechanism of "dipole moment self-induction," which is based on the Bernal–Fowler water model, assumes that the reference H_2O molecule is immersed in a medium of much smaller polarizable particles. However, in Bernal–Fowler water, molecules are all the same size, and the closing fields required by Onsager's theory cannot be built with identical dipoles [11]. The problem of Onsager's approach originates from the inappropriate mixing of macroscopic and microscopic concepts. The theory of the local field fails, because it does not account for intermolecular polarization effects, dealing with molecular dipoles only.

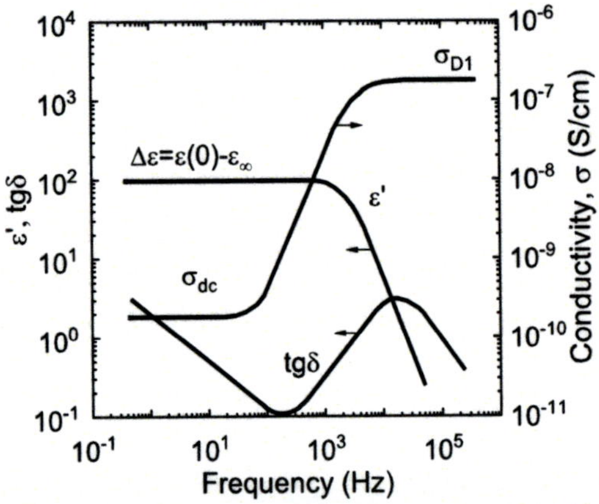

Fig. 2.11 The electrical properties: the dynamic conductivity (σ), the real part of the dielectric constant (ϵ'), and the loss tangent (tgδ = ϵ''/ϵ'), of a very pure ice crystal as a function of the frequency ν at -10 °C. Data from [37]

2.4.3 *Intermolecular Polarization Approach*

The alternative approach to the interpretation of the dielectric constant is based on accounting for intermolecular charge separation effects. Analyzing the frequency dependence of the conductivity of ice shown in Fig. 2.11, Gränicher [37] and Jaccard [38], suggested explaining the existence of two conductivity plateaus: the low frequency, σ_{dc}, and the high frequency, σ_{D1}, by ionized states (H_3O^+ and OH^-) and orientational L-D defects (doubly occupied or vacant bonds between the water molecules [39]), respectively. The ice spectrum shows two levels of conductivity, which differ by three orders of magnitude (see Fig. 2.11). The high-frequency plateau corresponds to Debye relaxation, and the entire spectrum can be described by the Debye formula, written in terms of dynamic conductivity:

$$\sigma - \sigma_{dc} = \frac{\sigma_{D1} - \sigma_{dc}}{1 + \frac{1}{i\omega\tau_r}}, \tag{2.51}$$

where the relaxation time, τ_r, corresponds to the transition frequency from the high- to low-frequency plateau, and σ_{D1} is defined through the static dielectric constant, $\epsilon(0)$, and τ_r by the following formula:

$$\sigma_{D1} - \sigma_{dc} \approx \sigma_{D1} = \frac{\varepsilon_0 \varepsilon(0)}{\tau_r}. \tag{2.52}$$

Equation (2.52) assumes that the static dielectric constant $\epsilon(0)$ is formed by the same mechanism as the high-frequency conductivity σ_{D1}.

Inasmuch as water and ice have much in common from the point of view of electrodynamics (see Chap. 3), and despite their dielectric relaxation times differing by seven orders of frequency magnitude, their spectra show a perfect scaling effect [17], which results, in particular, in the very similar dielectric constants. Similar spectra can be explained similarly, and the model of Gränicher and Jaccard can be extended to water. Thus, the static dielectric constant of both substances, following the Clauzius–Massotti approach, and using the Nernst–Einstein relation, $\sigma_{D1} = q^2/k_B T \cdot n_{\pm}$D, can be defined by

$$\varepsilon(0) - \varepsilon_{\infty} = \frac{q^2 n_{\pm} L^2}{6 k_B T \varepsilon_0}, \tag{2.53}$$

where $n_{\pm}$ is the concentration of charged defects, L and D are their separation distance and diffusion coefficient, respectively. The separation of L-D defects is diffusion controlled, and thus one can write for the mean-square displacement for the relaxation time:

$$L^2 = 6 D_{self} \tau_r. \tag{2.54}$$

Substituting (2.54) for (2.53), and applying the self-diffusion coefficient of water molecules, $D_{self} = 2.3 \cdot 10^{-9}$ m²/s (see Sect. 1.4), we get $n_{LD} = 6 \cdot 10^{27}$ m^{-3}. This value corresponds to 20% of all water molecules with a concentration of $n_w = 3.3 \cdot 10^{28}$ m^{-3}, which immediately implies a significant distortion of intermolecular interactions, and a strong interaction between charge carriers, including conventional H_3O^+ and OH^- ions.

The latter contradict the Bernal–Fowler model, in which intrinsic water ions have been assumed to be independent particles (see Sect. 1.2). Furthermore, two types of electrostatically interacting defects with different mobilities would lead to two relaxation times, which are not experimentally observed. In other words, Jaccard's theory fails to adequately reproduce the Debye relaxation and the dielectric constant of both ice and water.

In addition, the mechanism of polarization due to orientational defects is in principle similar to that considered by Onsager, because it reduces to the collective dynamics of perturbed H_2O dipoles, and, thus, has the same problems as the local-field approach discussed above (see Sect. 2.4.2). In particular, the relatively long relaxation time, τ_r, of water and ice assumes that no orientational defects could survive due to the relative molecular reorientations and translations, and that the dipole moment of hypothetical L-D pairs would be averaged without contributing to dielectric constant.

In order to resolve the problem of the intermolecular polarization effect, and of the dielectric constant, Artemov et al. [8, 17] introduced a model with only one type of defect, the conventional ionic species (H_3O^+ and OH^-). Unlike Bernal–Fowler concept, where the ionic species are considered independent, this model accounts for the electrostatic interaction between ions. As a result, the high-frequency conductivity, $\sigma_1(D)$, of both water and ice is assigned to the dynamics of the short-lived H_3O^+ and OH^- ions, and the low-frequency static conductivity, σ_{dc}, is explained by the long-lived (pH active) ions of the same type. In this way, the frequency dependence of the conductivity shown in Fig. 2.11 is explained by the single mechanism of the diffusion of intrinsic ions of water with an exponential lifetime distribution (see Sect. 4.5.3).

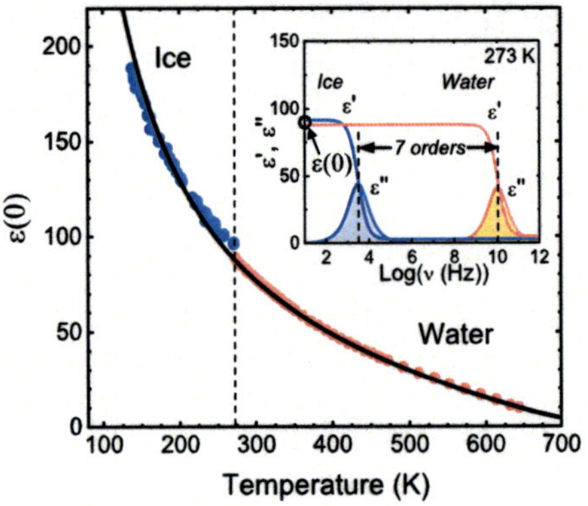

Fig. 2.12 The temperature dependence of the static dielectric permittivity of ice and water. The black solid line is the fit according to (2.55). The inset shows the real (ϵ') and the imaginary (ϵ") parts of the complex dielectric function of ice and water at 273 K. Reproduced from [8] with permission from the PCCP Owner Societies

The ionic model (see Sect. 4.2.3) shows that the relaxation time τ_r corresponds to the minimum time required for the percolation between ions (the static conductivity limit), and approximately equals the lifetime of H_2O molecules [8]. The static dielectric constant, $\epsilon(0)$, is explained in terms of the effective dipole moments formed by the separated H_3O^+ and OH^- ions, and is defined by

$$\varepsilon(0) - \varepsilon_\infty = \frac{q^2 n_i D_p \cdot \tau_r}{k_B T \varepsilon_0}, \tag{2.55}$$

where n_i is the concentration of H_3O^+ and OH^- ions, and $D_p = 9.3\times10^{-9}$ m^2/s is the diffusion coefficient of proton (see Sect. 1.4). Equation (2.55) is formally similar to (2.53), but the parameters have another meaning.

Figure 2.12 shows that the experimentally observed static dielectric constant satisfactorily fits (2.55) in the whole temperature range, providing an explanation for the static dielectric constant of water and ice for the whole temperature range on the same basis. The details of the ionic model are discussed in Chap. 4. Here we just note for clarity that the existence of a high concentration of excess protons $n_i \approx 10^{27}$ m^{-3}, which corresponds to the short-lived H_3O^+ and OH^- ions, has recently been independently confirmed by IR spectroscopy [40], and the simulation of molecular dynamics [41].

2.5 Infrared and Raman Spectra

Vibrational IR and Raman spectroscopy provide useful information about specific intra- and intermolecular dynamics on sub-picosecond timescales, and clearly show that water is not just an ensemble of individual H_2O molecules bound together. Here we focus on traditional methods of vibrational spectroscopy only, but there are many interesting non-linear spectroscopy and pump–probe techniques that have emerged over the past few decades, and which are reviewed, for example, in [42].

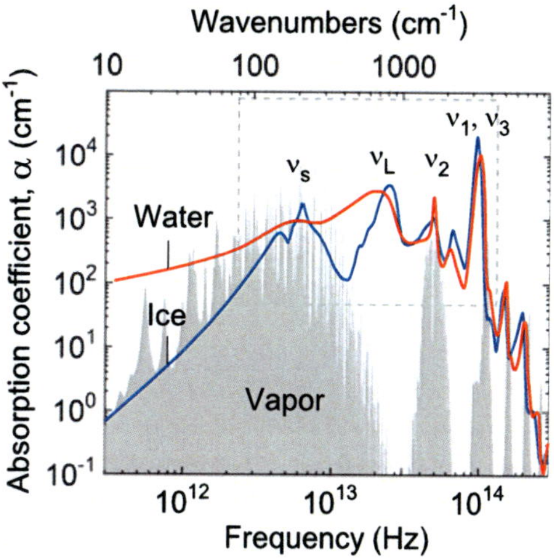

Fig. 2.13 The infrared absorption spectra of all phases of water: liquid (red), solid (blue), and gaseous (gray), in double-logarithmic scales. Frequencies ν_i (for i = 1–3) correspond to the fundamental vibrational modes of the water molecule. Frequency ν_s is $200\,\mathrm{cm}^{-1}$-mode, which is discussed in Sect. 2.6. The dashed area designates the part shown in Fig. 2.14 in linear scales

These methods complement the scope of spectroscopic experimental data on the electrodynamic properties of water in the IR frequency range, but lie outside of the current book.

We first discuss the vibration of H_2O molecule. The water molecule is an asymmetric top [43] that has rotational and vibrational modes, both effectively resulting in the absorption of electromagnetic radiation in the IR spectrum, which lasts from approximately $10^1\,\mathrm{cm}^{-1}$ to approximately $10^4\,\mathrm{cm}^{-1}$ (10^{12}–10^{15} Hz). The three fundamental vibrational modes of a single H_2O molecule are symmetric-stretch (ν_1), asymmetric-stretch (ν_3), and bending (ν_2) modes, which lie in the mid-IR region between 1,600 and 4,000 cm^{-1} (see Fig. 2.13). All three main modes are both IR and Raman active. The transitions between rotational levels are expected around $200\,\mathrm{cm}^{-1}$, and the corresponding lines of water vapor range from $1{,}000\,\mathrm{cm}^{-1}$ down to the microwave region. The real spectrum of water vapor also contains mixed vibrational-rotational transition line groups above $4{,}500\,\mathrm{cm}^{-1}$, whose intensity gradually decrease as the frequency increases, following Maxwell–Boltzmann statistics.[10]

While the absorption lines of water vapor can be predicted theoretically and identified (see, for example, HITRAN database,[11]), the spectrum of the condensed phases of water is still far from completely understood. Table 2.5 contains the frequencies of main maxima of the absorption bands of water in all aggregation states. One can see that spectra of water and ice differ from those for water vapor; the individual lines are overlapped and form broad absorption bands with a complex structure and an asymmetric shape (see Figs. 2.13–2.15). There are also several lines, which are

[10]The absorption of the electromagnetic waves by water vapor is used in IR astronomy and radio astronomy in the IR, and microwave or millimeter wave bands. For earth-based astrophysical observations, atmospheric water vapor creates distortions. The South Pole Telescope was constructed because there is very little water vapor in the atmosphere above the poles, caused by the low temperatures.

[11]https://hitran.org/.

Table 2.5 The wavenumbers in cm^{-1} corresponding to the maxima of the main infrared absorption lines of vapor, water, and ice H_2O and D_2O shown in Figs. 2.13 and 2.15 [46]

		ν_1	ν_3	ν_2	ν_L	ν_s
Vapor	H_2O	3657	3756	1595	–	–
	D_2O	2669	2788	1178	–	–
Water	H_2O	3400		1637	685	177
	D_2O	2500		1215	179	179
Ice	H_2O	3172		1648	825	181
	D_2O	2358		1226	No data	

absent in the vapor, such as the overtone between the OH-band and the ν_2 band, and the libration band, ν_L, near 600 cm^{-1}.

Two circumstances hinder the progress with the IR spectra of water: strong correlations and quantum effects. For instance, if two oscillators with the same characteristic frequencies and energies are connected, this leads to a coupling effect, which appears, for example, between modes ν_1 and ν_3. Due to the coupling, the line intensity and frequency can vary from those predicted from theoretical calculations. Another example of coupling is Fermi resonance. Due to the high cooperativity of water molecules, IR bands are wide and can overlap. For instance, the OH-stretching band, which lasts for several hundreds of wavenumbers, overlaps with the first overtone of the bending band, triggering a resonance between these two vibrations [44], which only affects the higher frequency band however. This effect, which is traditionally ignored in simulations, is important in the analysis of the fine structure of the OH-stretching band [45].

Fermi resonance and other coupling effects which affect the structure of IR spectrum of water can be excluded by isotopic substitution, the study of temperature dependencies, or the comparison of IR and Raman spectra. In the latter case, we analyze the transformation of the band shape as shown in Fig. 2.14. The main peak of the IR spectrum is at 3,400 cm^{-1} with a shoulder at 3,250 cm^{-1} and a width of 375 cm^{-1}, while the Raman spectrum has two peaks at 3,300 and 3,150 cm^{-1}, and a width of 340 cm^{-1}. Note that the vapor molecules have symmetric (ν_1) and asymmetric (ν_2) fundamental stretching modes, which are IR and Raman active at 3,657 and 3,756 cm^{-1}, respectively, and thus have a split about 100 cm^{-1} (see Table 2.5). However, the spectra of water are significantly redshifted from these values, and their breadths are much larger than the $\nu_1 - \nu_2$ split.

The interpretation of the IR and Raman spectra can be roughly divided into two approaches. The first one was suggested by Röntgen [47] and is based on the idea that water is a multicomponent mixture of discrete species. To treat the spectra one can use several Gaussians, which can then be attributed to a normal mode, or to vibrations in different molecular environments. For example, the OH-stretching mode can be split into several Gaussian peaks (from two to six), which can be assigned to molecular species with different coordination numbers [48–52]. The

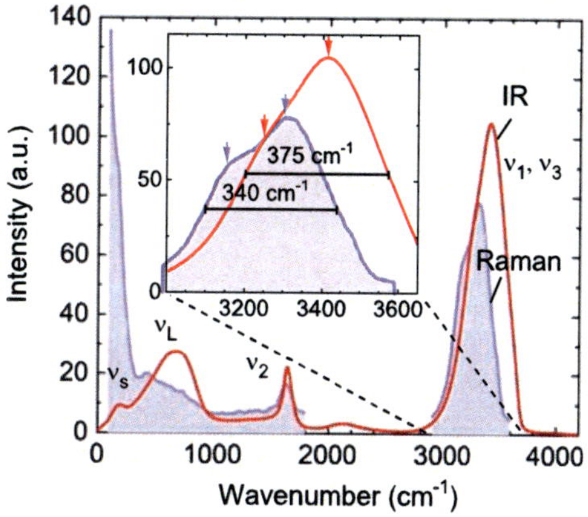

Fig. 2.14 Infrared absorption (red) and Raman (purple) spectra of water. The inset shows the enlarged OH-stretching vibration mode

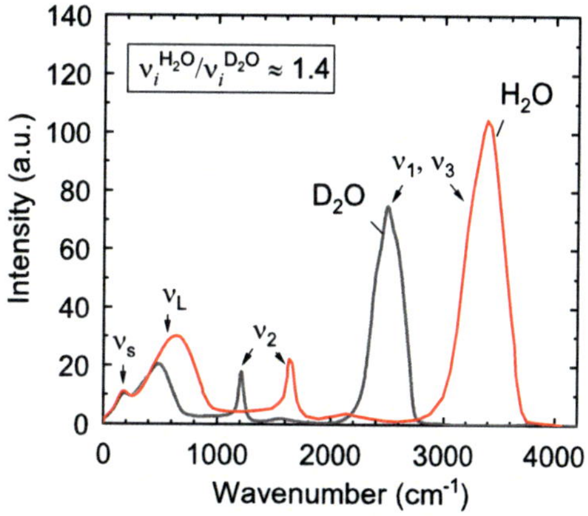

Fig. 2.15 The infrared absorption spectra of liquid H_2O (red) and D_2O (black). All modes shown are intramolecular, except the mode ν_s, which has been shown to be intermolecular (see the text)

results from such a procedure, together with the observation of isosbestic points on the temperature-dependent spectra, can be interpreted as supporting the mixture models of water [48, 49]. However, Geissler has shown [53] that a single solute species in a fluctuating environment exhibits isosbestic points without implying multiple species.

The second approach to the interpretation of the IR and Raman spectra is based on the continuum model. Water is considered as a continuum of several states with varying strengths of interactions between neighboring molecules [54]. The smooth transition between these states makes it impossible to separate the spectral lines into additive components [55]. The parallel comparison of the spectra of light and heavy water (see Sect. 2.7.2), and their mixtures, shows the monomodal shape of the OH-stretch vibration region, thus favoring continuum models of water [54]. Every description of water in terms of "broken" and "unbroken" bonds, or "monomers" and "clusters" is therefore arbitrary [56]. We avoid this approach in the further discussion.

Figure 2.15 shows the IR spectra of light and heavy water (more spectra can be found in Sect. 2.7.2). All absorption bands of D_2O, except ν_s, show a systematic redshift with the ratio $\nu_{H_2O}/\nu_{D_2O} \approx 1.41 = \sqrt{2}$, which corresponds to a change in the molecular torque when the hydrogen atoms (H) are replaced with twice-as-heavy deuterium atoms (D). The spectral bands of H_2O and D_2O are extremely broad, overlap, and have a highly complicated temperature dependence. However, the interpretation of the very complex bands of pure components can be done using the analysis of the spectra of semi-heavy water (HDO). The main advantage of these studies is that the three native fundamental vibrations of HDO, and their overtones and combination bands, are widely separated [56] so that Fermi resonance does not occur (at least for bending modes).

Maréchal [57] has measured the IR spectra of light (H_2O), heavy (D_2O), and semi-heavy (HDO) water, and found that the "pure" HDO spectra, obtained from the spectra of mixtures with different proportions of H_2O and D_2O, are slightly different near the bending modes (ν_2). Later Max and Chapados [58] showed, in a more detailed study, that mismatches are reproducible and consistently change with the molar fraction.

Recent work in this regard [40] showed that there are mismatches (see Fig. 2.16) in the IR spectra of HDO, obtained from different proportions of light and heavy water, which correspond to fingerprints of the fluctuation-born short-living H_3O^+, DH_2O^+, HD_2O^+, and D_3O^+ ions with concentrations of about 2% of the content of water molecules. These ions, in which the bending-mode model is shown in Fig. 2.17, presumably coexist with long-lived pH-active ions, thus making liquid water an effective ionic liquid on the picosecond timescale.

The origin of two far-IR absorption bands at $\nu_s = 200\,\text{cm}^{-1} = 5$ THz and $\nu_3 = 60\,\text{cm}^{-1} = 1.8$ THz (see also Fig. 2.4) is not clear at the moment. These modes show an intermolecular nature, as they do not show the isotope effect (see Sect. 2.7) and have high intensities, which assume unusually strong intermolecular interactions [59].[12] LO–TO splitting[13] of the mode near $240\,\text{cm}^{-1}$ was experimentally confirmed [60], which means that the long-range force effect exists in ice. The detailed structure and properties of the ν_s and ν_3 modes are still poorly studied. However, time-domain spectrometers, which appeared a decade ago, triggered the studies of the terahertz spectrum of water and ice, which is considered in Sect. 2.6.

Summarizing, the vibrationally averaged microscopic dynamics in water do not reduce to the oscillatory dynamics of H_2O molecules only. The strong correlations and the intermediate states of protonic transport affect the molecular modes and contribute to the IR spectrum. In particular, it was shown that Zundel and Eigen cations (see Sect. 1.3.2) contribute to the whole IR region [61–63]. Another important observation is that the IR spectra of ice and water are very similar. That automatically assumes a unified interpretation, which is missing at the moment. For example, a longitudinal and transverse dispersion resembling optical phonons has been demonstrated for ice [60], but was never discussed for water until recently [64]. Inasmuch as

[12]The maximum of the IR absorption of water and ice is only an order of magnitude lower than that for the ionic crystal of NaCl, but three orders of magnitude lower than the absorption of covalently bounded crystalline silicon.

[13]LO–TO splitting manifests itself in a frequency difference between the longitudinal optical (LO) and transverse optical (TO) phonon modes.

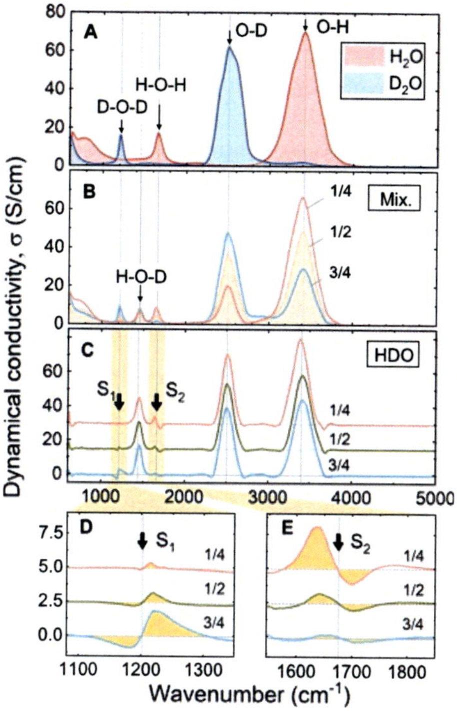

Fig. 2.16 The experimental spectra of the dynamic conductivity of **a** pure H_2O and D_2O, **b** their mixtures with different molar fractions, **c** pure HDO calculated by the subtraction of the pure spectra from the spectra of mixtures, and **d**, **e** magnified HDO spectra near the bending modes showing "S-shaped" mismatches S_1 and S_2 assigned to the short-lived excess-proton states (H_3O^{+*} and OH^{-*}). Reproduced from [40] with permission from Springer Nature

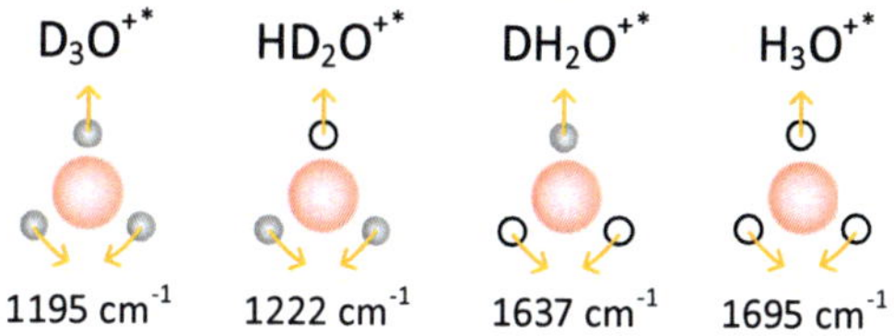

Fig. 2.17 The bending oscillations of positively charged ions in mixtures of light and heavy water with corresponding eigenfrequencies. Gray circles are D atoms, small open circles are H atoms, and large red circles are O atoms (see the text). Adapted from [40] with permission from Springer Nature

LO–TO splitting assumes long-range correlations, we should consider corresponding changes to the Bernal–Fowler water model, as the model does not assume any long-range order for water. Finally, the intense far-IR part of the water spectrum indicates strong molecular correlations. This part of the spectrum, insensitive to isotopic substitution, requires intermolecular correlations that go beyond the model of the network of bounded H_2O molecules.

2.6 The Terahertz Spectrum of Water

2.6.1 Between the Infrared and Dielectric Spectra

Vibrational spectroscopy studies intramolecular oscillations and dielectric spectroscopy is a reflection of collective dynamics, while terahertz spectroscopy, which lies at the frequencies between these two methods, probes the transition between oscillatory and diffusion motion in water and ice. That is why the continuous spectrum of water was missing until terahertz data appeared. Time-domain laser spectroscopy made the terahertz data widely spread; however, there are still some technical difficulties.

The first comprehensive collection of spectral data points was made by Segelstein in 1981 [69] and is still used for analysis. Mid-IR and far-IR data were measured by Zelsmann in 1995 [65] and updated by Bertie et al. in 1996 [70], where the attenuated total reflection (ATR) technique was used. A more recent collation of data was made by Artemov in 2014 [71], and new experimental data from 500 MHz to 400 THz were provided by Shiraga et al. in 2018 [72]. Despite the great work on data collection and the availability of many independent measurements, the data points still vary significantly depending on the sample geometry, the method used, and the data processing procedure.

Figure 2.18a collates spectral data points for water, which were obtained over the past few decades. The terahertz–infrared part of the spectrum is enlarged on the right-hand side of the graph. Although a generally good correlation between the data points has been achieved, there is still a large degree of uncertainty in the terahertz frequency region (mainly in the real part of the dielectric permittivity). In particular, there is uncertainty in both the peak center and integrated absorption cross section of the ν_s mode centered at about 5 THz (200 cm^{-1}), and there is a divergence of the data in between 0.1 and 1 THz, which contain both secondary relaxations: ν_{D2} and ν_{D3}.

The terahertz region covers the intermolecular vibrational modes, which appear in the frequency window from 1 THz up to about 10 THz, corresponding to 30–300 cm^{-1}. This part of the broadband spectrum reflects the dynamics in the first and the second molecular coordination spheres at distances of about 6–7 Å. The long-range sensitivity of terahertz absorption to the dynamic properties of water

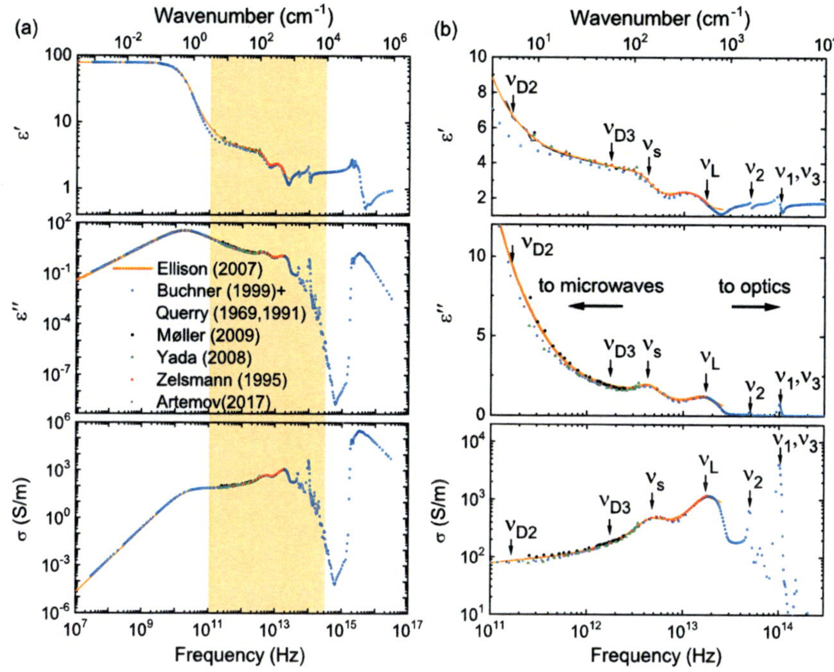

Fig. 2.18 A collection of experimental data points for water in terms of the dielectric permittivity (real, $\epsilon'(\omega)$, imaginary, $\epsilon''(\omega)$, parts), and dynamic conductivity, $\sigma(\omega)$: **a** the full spectral range, **b** the enlarged terahertz–infrared part. Arrows show the characteristic frequencies, discussed in Sects. 2.3 and 2.5. Data from [6, 7, 65–68] are used

results from processes in water occurring on picosecond timescales. However, a full interpretation of the terahertz spectroscopy data remains a challenging task.

2.6.2 5 THz Oscillation Mode

The oscillatory mode ν_s near 5 THz (200 cm^{-1}) (see Fig. 2.18b) of the far-IR spectrum of water has properties that differ from other IR oscillations observed at higher wavenumbers. In particular, it is conserved in isotopic substitution and phase transition [73], and, unlike higher frequency librational mode ν_L, near 600 cm^{-1}, is not properly reproduced by conventional molecular-dynamic simulations [74–77]. This mode does not correspond to any fundamental modes of water molecules or their combinations. Heyden et al. [75] referred to ν_s as collective intermolecular stretching vibrations, and others discuss a translation motion of molecules with intermolecular charge transfer induced by the OH-stretching mode [76, 77]. However, these interpretations do not account for mode ν_s disappearing in the supercritical state [78], and being weakly influenced by the presence of electrolytes [79]. The mode can be

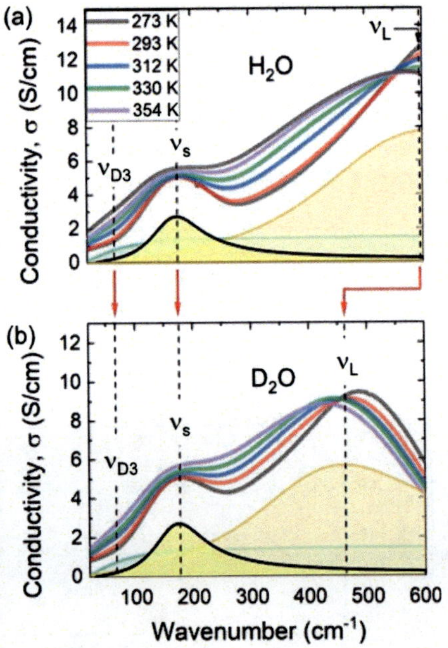

Fig. 2.19 The terahertz spectrum of **a** water and **b** heavy water near the 5 THz oscillation mode, ν_S. The mode does not change with temperature or isotopic substitution. Experimental data by Zelsmann [65]

fitted as a separate oscillator from the relaxation part of the spectrum (see Sect. 2.3.1) which is contrary to its observed isotopic behavior (see Sect. 2.7), or it can be considered as part of the uniform intermolecular dynamics, which extends to the dielectric relaxation part of the spectrum [8].

Figure 2.19 shows the terahertz part of the dynamic conductivity spectrum of water and heavy water at different temperatures. The spectral region, besides the small relaxation part below 150 cm^{-1}, contains two oscillatory modes: ν_L near 600 cm^{-1} and ν_s near 200 cm^{-1}, which was mentioned above. The former shifts when the hydrogen atoms are replaced by deuterium atoms, while the latter is conserved. When the temperature decreases, the mode ν_L shows a redshift, while the mode ν_s stays at the same position, and seems stable even in the solid state (see Fig. 2.13).[14]

Table 2.6 shows the parameters of the mode ν_s for H_2O, D_2O, H_2O^{18}, and ice. The central frequencies and the intensities, which are represented by their dielectric contributions $\Delta\epsilon$, coincide within a few percent for all the substances. For water, the damping γ_s is approximately equal to the central frequency ν_s, which means that the corresponding oscillatory motion is overdamped, due to the rapid rearrangement of the environment of the oscillating particles. For ice, on the contrary, a smaller value of γ_s in comparison with that for ν_s indicates a long-lived oscillatory state.

Figure 2.20a compares the Raman spectrum and the vibrational IR spectrum of water, which look very similar to each other. The Raman spectrum has two oscillatory modes, ν_{D3}^R near 60 cm^{-1} ($\approx$1.8 THz) and ν_s^R near 175 cm^{-1} ($\approx$5.3 THz), which are

[14]For ice, the ν_s mode splits and has a structure of at minimum two components, which is presumably caused by the longitudinal- and transverse-phonon modes splitting.

Table 2.6 Best-fit parameters of the oscillatory 5 THz mode of liquid H_2O, D_2O, and H_2O^{18} at room temperature (data from [80]) and ice at 266 K

	$\Delta\epsilon_s$	ν_s (THz)	γ_s (THz)	ϵ_∞
H_2O	1.25	1.11	1.22	0.85
D_2O	5.30 (0.05)	5.36 (0.03)	4.95 (0.04)	5.44 (0.03)
H_2O^{18}	5.40 (0.20)	5.06 (0.08)	4.25 (0.09)	2.93 (0.25)
Ice	2.34 (0.02)	2.29 (0.01)	2.28 (0.01)	2.2 (0.1)

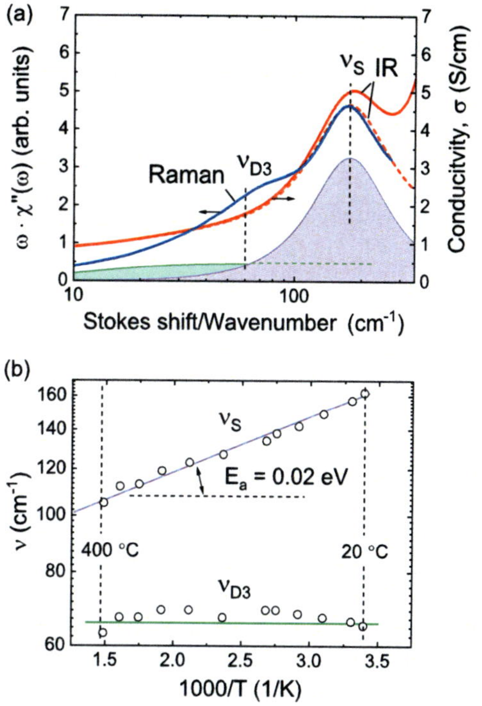

Fig. 2.20 **a** A comparison of the Raman [81] and infrared spectra of water. The dashed line is obtained by subtracting the high-frequency modes. The shaded areas are the components of the spectrum according to (2.25) and (2.26). **b** The temperature dependence of the Raman mode central frequencies. Calculated using data from [82]

close to those in the IR spectrum. The ν^R_{D3} corresponds to the third relaxation mode of the IR spectrum (see Fig. 2.4) and the ν^R_s coincides with the ν_s mode of the IR spectrum, discussed above. Note that the IR-ν_s mode is perfectly matched with that in the Raman spectrum when the high-frequency contributions are taken out (see the red dashed line Fig. 2.20a).

Two main features of the Raman spectrum, ν^R_{D3} and ν^R_s, have been assigned, by analogy with the IR spectral features, to O–O–O bending [83], and O–O stretching along the O–H···O line [84], respectively. Walrafen showed [85] that the 60 and 175 cm^{-1} Raman peaks have a more complex nature, and correspond to transverse spherical acoustic shear and longitudinal spherical acoustic dilatational waves, respectively, and both involve the displacement of the center of mass of H_2O, which

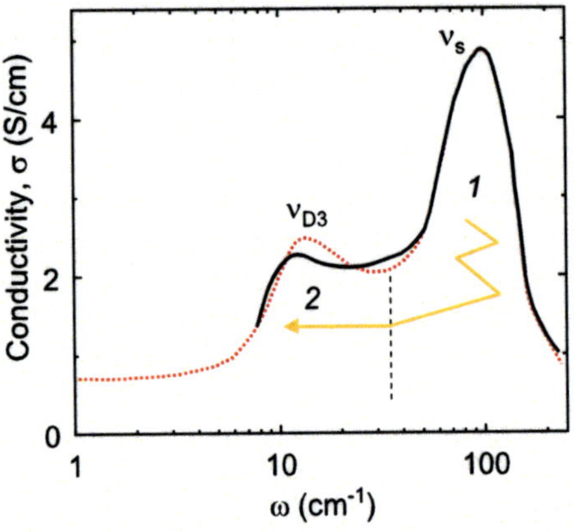

Fig. 2.21 The Raman spectrum of the superionic conductor α-AgI. Modes ν_{D3} and ν_s are marked by analogy with those for water (see Fig. 2.20). The dashed line is the model which considers coupled motions of mobile ions and cage ions (see Sect. 3.5 for the same model applied to water). The numbers correspond to: (1) the oscillatory motion of a "dressed" ion and (2) the bare diffusion of ion to the nearest "cage." Data from [86]

is possible through the interstitial mechanism discussed in Chap. 1.[15] The latter explains the insensitivity of both peaks to isotopic substitution, because the oxygen atom is an order of magnitude heavier than the hydrogen/deuterium atom, and thus displacement does not change the molecular weight significantly. Figure 2.20b shows the temperature dependencies of the central frequencies of the modes ν_{D3}^R and ν_s^R. The former is temperature independent, while the latter shows a redshift as the temperature increases with a very small activation energy of 0.02 eV, which is close to $k_B T$.

The described spectral features and their behavior are similar to those observed in superionic conductors, dielectric materials with high ionic conductivity, whose structure is characterized by disorder in the sublattice of the conducting ions. For example, Fig. 2.21 shows the typical Raman spectrum of silver iodide (AgI). The large peak belongs to vibrations of the silver atom near the equilibrium position, while the small one is connected with the diffusion of the same atom from one quasi-equilibrium state to another [87]. Comparing Figs. 2.21 and 2.20, one can see that the spectra are qualitatively similar to each other, and even have comparable intensity of their peaks.

The models of electrical conductivity of superionic conductors were developed in 1970s, when the lattice-gas model was used for the explanation of Raman spectrum similar to that shown in Fig. 2.21 [87]. In case of AgI, the I^- ions are considered to form the lattice, while Ag^+ ions are mobile and form a kind of fluid in the frame of reference of the stable iodide lattice.

[15] Note that the acoustic waves involve the motions of entire H_2O molecule and describe the irregular molecular arrangement, whereas the X-ray RDF analysis (see Sect. 1.2.3) gives diffusion-averaged O–O distances. That is why an additional small maximum of the radial distribution function near 3.5 Å can be caused by the molecules in the state of diffusion between two quasi-equilibrium positions.

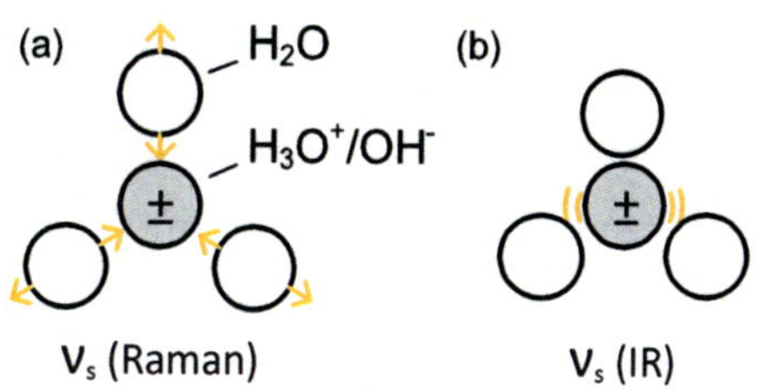

Fig. 2.22 Possible molecular dynamics that correspond to the ν_s mode in the **a** Raman and **b** infrared (IR) spectra of water

Following the analogy with superionics, the oxygen atoms seem to play the role of the "lattice," while the hydrogen nuclei (protons) form a disordered fluid (or quasi-gas of protons and proton holes) in the frame of reference of the oxygen atoms (or water molecules). The redshift of the AgI spectrum with respect to the spectra of water can therefore be explained by the difference in the molecular masses of the silver ion (M_{Ag} = 108) and water ions ($M_{H_3O^+,OH^-}$ = 19 or 17). The latter are considered, because unlike the AgI, where Ag ions are free to move, the protons of water are always attached to the oxygen atoms (see Sect. 1.4). The good coincidence of the spectral shapes of water and superionic conductors in the far-IR region opens up the possibility for the further development of the microscopic model of water by analogy with superionics.

Decka et al. [79] discuss terahertz fingerprints of the solvated proton and show that it has two resonances, which they assigned to a solvation water mode (or the H_3O^+ rattling mode) of around 140 cm^{-1} and a blueshifted hindered translational mode of an Eigen species H_3O^+ in the surrounding water molecules (see Fig. 2.22b) at 325 cm^{-1}. This fact is in agreement with the result by Lapid et al. [88] who found oscillatory motion of the excess proton in a H_3O^+ moiety and its nearest neighbor oxygen prior and after a proton transfer event. The same mode was obtained by Kim et al. [89] in the vibrational density of states based on multistate empirical valence bond simulations. Interestingly, Eigen [90] considered the excess-proton delocalization within the H_9O^+ Eigen complex as a very fast process on the sub-picosecond timescale, which corresponds to the terahertz frequency range.

Thus, a possible explanation of the mode ν_s is a rattling of H_3O^+ ions. The possible mechanism of the dynamics behind the terahertz spectrum of water is discussed in [73]. The dielectric contribution $\Delta\epsilon_s$ of the mode ν_s allows one to estimate the concentration $n_\pm$ of ions. For the harmonic oscillator approximation:

$$n_\pm = \frac{\Delta\varepsilon\varepsilon_0(2\pi\nu_S)^2 m^*}{q^2}, \tag{2.56}$$

where q and m^* are the charge and effective mass of the ion, respectively. Equation (2.56) gives $n_\pm \approx 10^{27}$ m^{-3}, which corresponds to a few percent of all water molecules, thus providing a significantly higher value, which is usually used in the concept of pH. Nevertheless, as discussed in Chap. 4, short-lived excess-proton states can explain the majority of the spectral features of the broadband dielectric spectrum of water.

2.6.3 Second Dielectric Relaxation ("Excess Wing")

The frequency gap between IR oscillations and the dielectric relaxation is the most poorly studied part of the dielectric response of water. The corresponding frequency range from 50 GHz to about 10 THz is hard to reach by dielectric spectroscopy techniques from the left side and IR spectroscopy from the right side. That is why, the right-hand side of Debye relaxation has been long unstudied, and even today a large uncertainty of the data point exists, as shown in Fig. 2.18b, in this part of the spectrum. Recently, terahertz time-domain spectroscopy revealed two sub-picosecond spectral features, which have been treated as fast relaxation processes, ν_{D2}, near 60 cm^{-1} (2 THz), and ν_{D3}, near 5 cm^{-1} (0.2 THz) [91], which constitute the so-called "excess wing" of the dielectric relaxation, because the dielectric losses here exceed the level expected from the Debye formula 2.24. This deviation from Debye's prediction has been called a "fast relaxation" process [9].

The relaxation process ν_{D2} was initially assigned to the reorientation of an individual water molecule [91], as predicted by computer simulations [92, 93]. This interpretation assumes a two-component structural model of water, which has been revealed by Raman spectroscopy [94] and neutron scattering [95]. However, a different interpretation has been proposed based on the results of ultrafast spectroscopy [96]. It has been shown that both the main and the fast relaxation processes are due to collective dynamics, in which the main relaxation is caused by the collective reorganization of water molecules [97], and the fast relaxation is the inertial motion of individual water molecules [98].

Yada et al. [7], using terahertz time-domain spectroscopy, showed that the temperature dependence and the isotope shift of fast relaxation are in good agreement with the individual relaxation mode, thus confirming the existence of free molecules in the sub-picosecond timescale. The fast relaxation ν_{D2} is assigned to the collision process, which means that the heterogeneity of the water is dynamic, and the ambient water is regarded as two-component mixtures in a sub-picosecond time interval.

The mode ν_{D3}, which has been shown to be Raman active and similar to those observed in superionic conductors (see Sect. 2.6.1) and by neutron scattering [99], is assigned to the hindered translations [100, 101]. Arbe et al. [99] compared the dielectric relaxation data and the neutron scattering data and found that on sub-picosecond timescales, the hydrogen atom moves in a "cage" with a size of 0.5 Å. Such a size corresponds to the amplitude of the vibration of hydrogen atoms (presumably together with the host molecule) and is much smaller than the intermolecular distance, which is 2.8 Å. These results confirm that the mode ν_{D3} is related to the local translational movements of molecules, presumably in the excess-proton state, as discussed in [73].

Several phenomenological models have recently been suggested to describe the main dielectric relaxation near 20 GHz and the high-frequency processes shown in Fig. 2.4) on the same footing. Figure 2.23 compares these models with the experimental data in terms of dielectric permittivity. The classic Debye model (magenta)

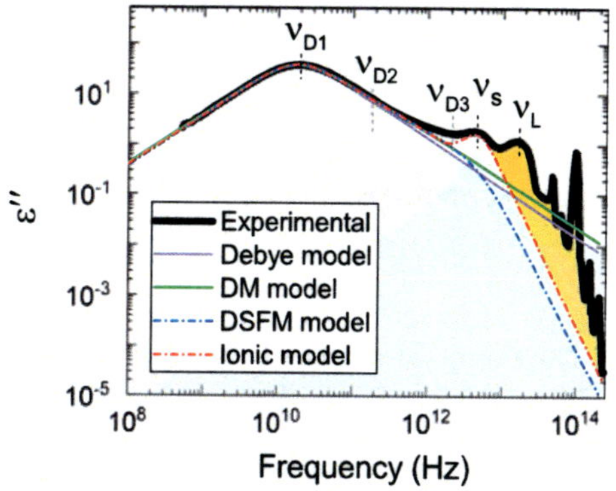

Fig. 2.23 The spectrum of the imaginary part dielectric constant of water. Colored lines are the Debye model (magenta), the DM model (green), the DSFM model (blue), and the ionic model (red). The shaded yellow area corresponds to the contribution of the vibration modes of water (see Sect. 2.7.2). See text for details

is valid up to about 80 GHz, and does not reproduce the high-frequency processes.[16] Popov et al. [19] introduced the defect-migration (DM) model (green), and assumed that the high-frequency dynamics, and the main relaxation, can be explained by the migration of orientational defects through the network of water molecules. However, the DM mechanism, as the Debye model, does not fulfil the sum rule (both models lead to the infinite dielectric loss paradox when the frequency tends to infinity) and has other problems discussed in Sect. 2.3.2. Figure 2.23 shows the Debye and DM models exceed the experimental dielectric losses in IR region.

Shiraga et al. [72] reconsidered the relaxational and vibrational line shapes, and suggested a discreet stochastic frequency modulation (DSFM) model (blue), which assumes the instantaneous modification of the line shapes by the correlation with the surrounding system. This model solved the problem of the conductivity sum rule (see Sect. 2.8) and reproduced the experimental dielectric spectra up to 1 THz (see Fig. 2.23), but did not account for the vibration mode ν_s, which is definitely a part of the intermolecular dynamics. Moreover, the DSFM model does not contain static DC conductivity.

An alternative model has been suggested by the author [8]. A fit within an "ionic model" of water is shown in Fig. 2.23 as a red line. One can see that the ionic model describes all the intermolecular spectral regions up to 10 THz, satisfies the sum rule, contains static DC conductivity (out of the graph), and describes water and ice on the same footing. According to this model, ice and water conduct electricity by the interaction of excess protons and proton holes. No other defects, except short-living H_3O^+ and OH^- ions, are required for the dielectric-spectra interpretation from DC up to terahertz. Within the model, the mode ν_s corresponds to the oscillatory motion of short-lived ionic species, the mode ν_{D3} is a free-pass translational motion of a newly appeared ion, the mode ν_{D2} corresponds to the solvation (hydration) of ions,

[16]Note that the Debye model of relaxation is based on the diffusion limit, where extremely complex dipolar relaxations at long timescales can be treated statistically, neglecting the moment of inertia of the dipoles and intermolecular interactions. However, for highly anisotropic and strongly dipolar molecules such as water, their effects should be rigorously taken into account at such short time frames as an equilibrium statistical description is inapplicable [102].

and the main relaxation band ν_{D1} is a result of the finite lifetime of ionic species (dielectric losses go down when the period of the external electric field exceeds the lifetime of the ionic species). The details of the ionic model are discussed in Chap. 4.

2.7 Heavy Water: H/D Isotope Effect

The atomic-molecular dynamics of water can be effectively tested by isotopic substitution. This assumes the replacement of one or several atoms by their isotopes, and a subsequent analysis of the transformation of the dielectric response. The simplest way to analyze the isotope effect on the spectrum is to compare the dielectric response of light (H_2O) and heavy (D_2O) water. Below we discuss the isotope effect in the relaxation (dielectric spectrum) and oscillation (IR spectrum) parts of the dielectric response or water.

2.7.1 Dielectric Spectrum: Intermolecular Dynamics

Figure 2.24 shows the dielectric spectra of ordinary and heavy water at room temperature in terms of the dielectric permittivity and dynamic conductivity. The spectra parameters are given in Table 2.7 at room temperature, and the coefficient of their temperature dependencies are given in Table 2.8. Up to 10^{13} Hz, the spectra show a minor response to the substitution of hydrogen by deuterium: the ratio of the main parameters of Debye relaxation is close to 1. The dielectric contributions of the main

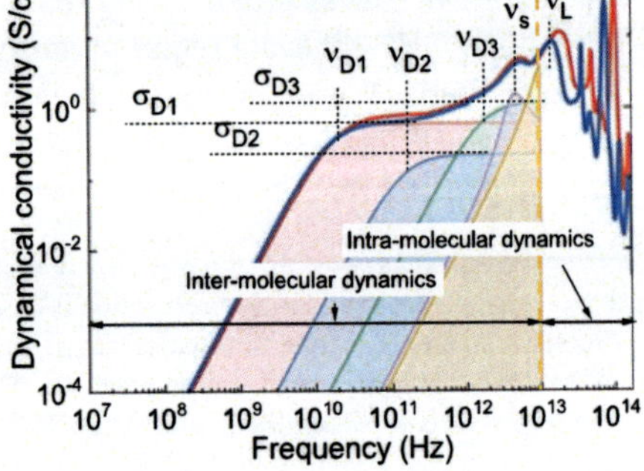

Fig. 2.24 The broadband dielectric spectra of ordinary (red) and heavy (blue) water at room temperature: the real and imaginary parts of the dielectric constant (top panel) and dynamic conductivity (bottom panel). They are data from [12]. The vertical line separates the regions of the strong and weak isotope effect (the relative shift of the spectra). Thin lines are fit components according to (2.26). The right part is shown separately in Fig. 2.30

Table 2.7 The best-fit parameters of the dielectric spectrum of H_2O and D_2O at room temperature according to (2.26). Frequencies are in terahertz and conductivity values are in S/m

	$\Delta\epsilon_{D1}$	$\Delta\epsilon_{D2}$	σ_{dc}	σ_{D1}	σ_{D2}	σ_{D3}	ν_{D1}	ν_{D2}	ν_{D3}	ν_s
H_2O	74.1	4.7	$5.5{\cdot}10^{-6}$	74	27	148	0.019	0.17	2.0	4.6
D_2O	74.9	4.8	$1.3{\cdot}10^{-6}$	66	53	125	0.016	0.16	1.8	4.6
Ratio	≈1	≈1	4.2	1.1	1.1	1.2	1.18	1.06	1.06	≈1

Table 2.8 The coefficients of temperature dependencies of the dielectric spectra of light (H_2O) and heavy (D_2O) water. According to the Arrhenius formula, $A(T)=A_0{\cdot}\exp(\Delta E/k_BT)$, where A_0 and ΔE are the pre-exponential factor and activation energy, respectively

	$A(T)$	$\Delta\epsilon_{D1}$	τ_{D1} (ps)	σ_{dc} (S/m)	σ_{D1} (S/m)
H_2O	A_0	14.0	0.41	7.9	$1.6{\cdot}10^4$
	ΔE	0.042	0.190	–0.370	–0.14
D_2O	A_0	13.9	0.69	6.6	$1.1{\cdot}10^4$
	ΔE	0.042	0.188	–0.398	–0.15

(Debye) relaxations, $\Delta\epsilon_{D1}$, almost coincide, and the small redshift of its relaxation frequency, ν_{D1}, appears, accompanying the decrease of the microwave conductivity plateau, σ_{D1}. The second and third relaxations follow the main relaxation and show a slightly bigger redshift of the central frequencies ν_{D2} and ν_{D3} compared to the main relaxation. However, the much smaller amplitude of both secondary relaxations assumes a large uncertainty in the corresponding parameters.

The temperature dependencies of the main parameters of the relaxation spectrum are collated in Fig. 2.25. The coefficients of the temperature dependencies of the high-frequency conductivity, σ_{D1}, the relaxation time, τ_{D1}, and the static dielectric constant $\epsilon(0)$ are obtained by the Arrhenius formula $A(T)=A_0{\cdot}\exp(\Delta E/k_BT)$, and the best-fit parameters are given in Table 2.8. The magnitude and the activation energy values of heavy water are close to those for ordinary water for all the parameters, which indicates that the potential barriers are of the same amplitude. The static dielectric constants, $\epsilon(0)$, of H_2O and D_2O coincide over a wide temperature range with an accuracy better than 3% (see Fig. 2.25c), thus differing from the behavior prescribed by the rotational polarization mechanism. The ratio between the relaxation times is about 1.2 for light and heavy water, and about 1.5 for ice (see Table 2.7). The general properties of the main relaxation shows that either the role of rotation molecular diffusion in the relaxation process is minor or negligible.

Okada et al. [12] found that the static dielectric constants of D_2O and H_2O follow the polynomial formula introduced by Uematsu [28], in which the number density instead of the mass density is used as the input parameter. The dielectric relaxation time decreases rapidly with increasing temperature and, unusually, jumps to a very large value at the liquid–gas transition. This result reveals that water molecules cannot be treated as point dipolar particles because the dielectric relaxation time of an isolated point dipolar particle would be very short. The authors measured

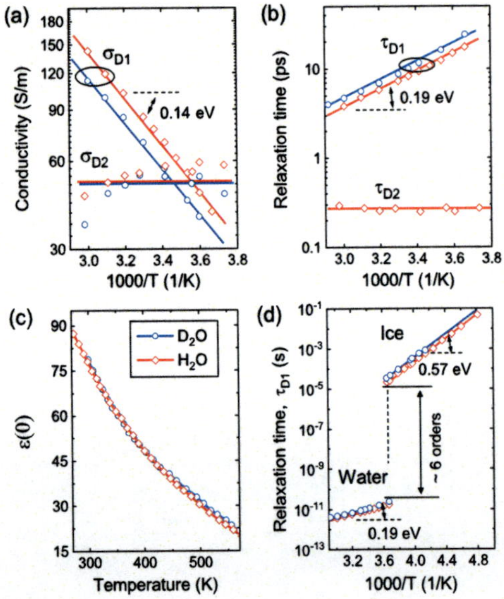

Fig. 2.25 The temperature dependencies of **a** high-frequency conductivities, **b** first and second relaxation times, **c** static dielectric constant, and **d** the first relaxation times of the liquid and solid states of light (H_2O) and heavy (D_2O) water. The data are from [12, 103, 104]. Numbers near curves are activation energies. The best-fit parameters are in Table 2.8

the dielectric relaxation under a DC bias field of 30 kV/m and confirmed that this strong field does not influence the relaxation time, τ_{D1}, which one would expect for the molecular dipole reorientation mechanism. They concluded that dielectric relaxation in the gaseous state is governed by the binary collision of water molecules, and explained the relaxation time quantitatively by introducing the collision time[17]:

$$\tau_c = \frac{1}{4\pi\rho r d^2}\sqrt{\frac{m\pi}{k_B T}}, \tag{2.57}$$

where m is the mass of molecule, ρ is the density, and d is the diameter of the water molecule. The authors showed that for dense water vapor the relaxation time, τ_r, is equal to the collision time, defined by (2.57).

Figure 2.26 shows separate light and heavy water molecules, which are asymmetric tops with three different moments of inertia along one of three axes: a, b, or c. The basis of such molecules is a relatively heavy oxygen atom (O) with a molecular weight of 16, and two relatively light hydrogen atoms (H) with the molecular weight of 1. Note that latter are only about 5% of the total molecular weight. The center of mass of the isolated water molecule is close to the center of the oxygen atom (see black dots in the figure), and the angular momentum is formed mainly by the hydrogen or deuterium atoms.

Table 2.9 shows the eigenfrequencies and rotational constants A, B, $C = h^2/(8\pi^2 I_{A/B/C})$, where $I_{A/B/C}$ are angular moments, for ordinary and heavy

[17] They also found that the product $\tau_{D1} \cdot D$, where D is the self-diffusion coefficient of water, is nearly independent of both light and heavy water

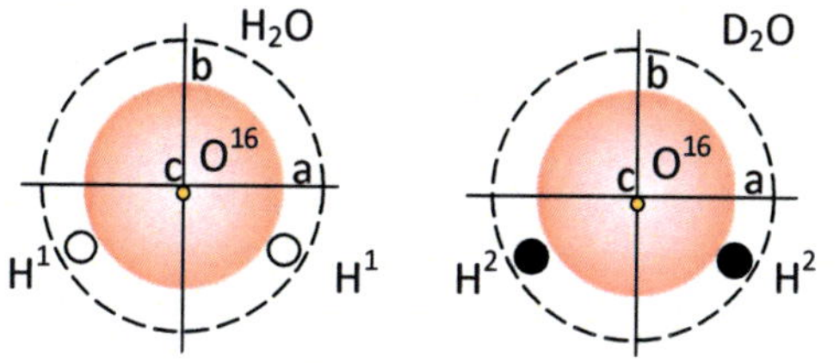

Fig. 2.26 The scheme of an H_2O molecule (left) and a D_2O molecule (right). Short-dotted circles indicates the molecular border with a diameter of 2.8 Å according to X-ray data. Hydrogen and deuterium atoms are shown by small open and close circles, respectively, both lie on the distance of the core-electron orbit radius from the center of the oxygen atom (O^{16})

Table 2.9 The rotational constants and eigenfrequencies of free light and heavy water molecules, according to [43]. Rotational constants are given in terahertz, and frequencies are in cm^{-1}

	A	B	C	ν_1	ν_2	ν_3
H_2O	0.8332	0.4347	0.2985	3694	1615	3802
D_2O	0.4615	0.2177	1.455	2666	1178	2787
Ratio	1.8	2.0	1.92	1.4	1.4	1.4

water molecules, shown in Fig. 2.26. The ratio between rotational momenta for D_2O and H_2O is about 2 regardless of the axis of rotation, and the ratio of the eigenfrequencies is close to ($\sqrt{2}$), because in the frame of reference of the heavier atom, the ratio of eigenfrequencies between H_2O and D_2O is determined by the ratio of the masses of protons and deuterons by the following formula:

$$\sqrt{\frac{m_1}{m_2}} = \sqrt{\frac{m_D}{m_H}} = 1.41, \quad (2.58)$$

where m_H and m_D are the masses of the proton and the deuteron, respectively. In other words, for free rotations, the moments of inertia are determined by the hydrogen (H) and deuterium (D) atoms only.

For the translational motion of molecules, shown in Fig. 2.26, the moment of inertia is formed by the whole molecule, and the ratio of eigenfrequencies is proportional to

$$\sqrt{\frac{m_1}{m_2}} = \sqrt{\frac{m_{D_2O}}{m_{H_2O}}} = 1.05, \quad (2.59)$$

where m_{D_2O} and m_{H_2O} are the masses of the heavy and light water molecules, respectively. This value is close to unity, and almost 1.5 times lower than that obtained by (2.58).

Yada et al. [7] showed that the collision time defined by (2.57) corresponds to the second relaxation, ν_{D2}. Using the Debye formula (2.38) and (2.57), and assuming $\tau_c = \tau_{D2}$, one gets

$$\frac{\sigma_{D2}^{H_2O}}{\sigma_{D2}^{D_2O}} = \frac{\rho_{D_2O}}{\rho_{H_2O}}\sqrt{\frac{m_{H_2O}}{m_{D_2O}}} = \frac{1.11}{1.05} \approx 1.1, \qquad (2.60)$$

which is close to the experimentally observed value (see Table 2.7), and close to the value prescribed by (2.59).

Thus, one can conclude that the gas-phase relation time, τ_r, introduced by Uematsu (2.57), corresponds to the second relaxation time, ν_{D2}, of liquid water, and does not directly correlate with the main Debye relaxation, ν_{D1}. The fact that the ratio $\nu_{D1}^{D_2O}/\nu_{D1}^{H_2O} \approx 1.18$ (see Table 2.7) is closer to that obtained by (2.59) than that by (2.58) indicates that the mechanism of the main relaxation of water, and the dielectric constant formation, is more like collective translational dynamics than collective reorientations. Since the conductivity plateau σ_{D1}, which is part of the main relaxation band, corresponds to the intermolecular charge dynamics (see Sect. 2.3), it is reasonable to associate these charges with excess protons/proton holes, which are apparently in the form of H_3O^+ and OH^- ions.

2.7.2 *Infrared Spectrum: Intramolecular Dynamics*

Figure 2.27 shows the IR region of the dynamic conductivity spectra of light and heavy water in different representations: (a) linear scales, (b) linear-logarithmic scale, and (c) double-logarithmic scales. Table 2.10 shows the frequencies of the maxima and integrals of the main peaks. The figure and table show that the spectra undergo a systematic shift with isotopic substitution, except the region of the peak ν_s, which is discussed in Sect. 2.6.1. The frequency shift of all peaks above 300 cm^{-1} occurs in accordance with formula (2.58), which assumes the intramolecular nature of the corresponding dynamics. In other words, peaks correspond to the dynamics of protons/deuterons in the frame of reference of the parent oxygen atoms. On the contrary, the lowest frequency resonance ν_s does not exhibit an isotopic effect, and, thus, should be attributed to intermolecular vibrations or the dynamics of excess charge together with the translation of the oxygen atom.

The mode $\nu_1+\nu_3$ is commonly agreed to be the O–H and O–D stretching vibrations. Their ratio $\nu_{1,3}^{H_2O}/\nu_{1,3}^{D_2O}$ is equal to 1.35, which is close to that expected by formula (2.58). However, this mode is not a single Loretzian and has considerable substructure, which is different for H_2O and D_2O [105]. In particular, the O–D band of D_2O has three distinct peaks at 2,395, 2,479, and 2,587 cm^{-1}, while H_2O has a similar structure but the peaks are far less distinct and the high-frequency peak has essentially disappeared [106]. The most common interpretation of the fine structure of this mode is based on the idea that there are several types of interactions between molecules, which influence the relative vibrations of hydrogen and oxygen atoms [54, 107]. Within this approach, the low-frequency shoulder has been assigned to an overtone of the bending mode, enhanced by Fermi resonance [108, 109] or, alternatively, to strongly bonded molecules in structured geometries [110, 111]. The

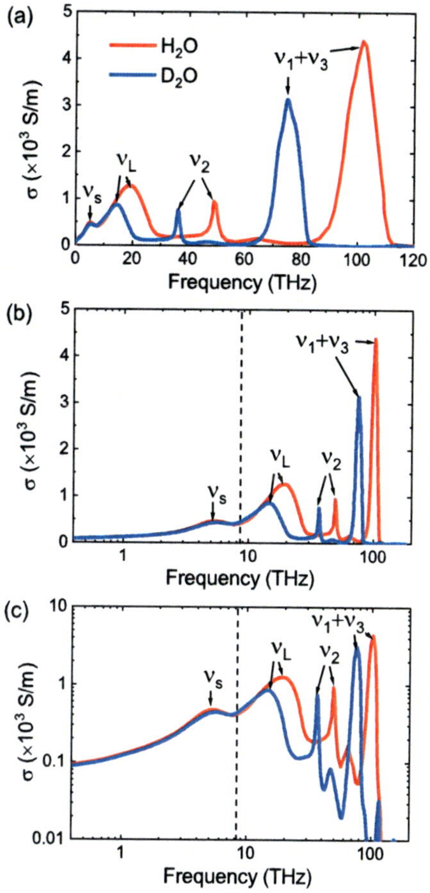

Fig. 2.27 The infrared part of the dynamic conductivity spectrum of light and heavy water in: **a** normal, **b** logarithmic and **c** double-logarithmic scales. The vertical dashed line separates the intramolecular and intermolecular modes

differences of these spectra for light and heavy water also indicate that there are nontrivial isotopic effects present.

Maréchal [105], analyzing the structure of the IR spectrum of water, found that the band ν_2, which is commonly assigned to molecule bending, has a perfect single-Lorentzian shape for light and heavy water. This fact excludes several types of interactions, which was assumed from the analysis of the O–H stretch band $\nu_1+\nu_3$. Maréchal concluded that water is made of H_2O molecules which either perform librations around one of their three axes (see Fig. 2.26) or perform rotations around an axis close to their C_2 symmetry axis, b, of a diffusional (relaxational) type. Inasmuch as molecules are distributed by energy, their rotational energy levels are either below or above the potential barrier of rotation. The molecules of the first type perform librations around their three axes, while molecules of the second type can rotate around the b-axis, but keep vibrating along two other axes, a and c. The latter molecules make water fluid. Such a model satisfactorily explains the properties of the mode ν_2, including its isotopic shift.

Table 2.10 Frequencies, ν, in Hertz and areas S in 10^{16} S/m·Hz of the infrared peaks of the dynamic conductivity spectra of light and heavy water shown in Fig. 2.27

	ν_s	ν_L	ν_2	$\nu_1+\nu_3$	S_s	S_L	S_2	$S_{1,3}$
H_2O	5.3	19.0	49	100	0.047	2.0	0.6	5.3
D_2O	5.2	14.5	33	75	0.047	1.0	0.3	2.7
Ratio	1.0	1.3	1.4	1.35	1.0	2.0	2.0	2.0

We conclude that the isotope effect analysis reveals that the spectra above 300 cm^{-1} (or about 10 THz) are mainly due to intramolecular dynamics, though some minor effects, such as collective phonon modes and quantum dynamics, also contribute and must be accounted for. The spectra below 300 cm^{-1} are, on the contrary, due to intermolecular, essentially collective, dynamics, which start from the oscillatory mode near 200 cm^{-1} and then manifest themselves as a broad relaxation band near 18 GHz with two high-frequency satellites, caused by molecular collision-rate-controlled dynamics.

2.8 The Conductivity Sum Rule

According to statistical mechanics, the complete conductivity tensor for a given frequency of the external electric field can be rigorously expressed in terms of electric current components fluctuating spontaneously in the equilibrium state [15]. The analytical properties of the corresponding real part of the dynamic conductivity function, $\sigma(\omega)$, allows one to introduce the so-called sum rule [1, 15]:

$$\frac{m}{2\pi^2\varepsilon_0 q^2}\int_0^\infty \sigma(\omega)d\omega = \int_0^\infty f(\omega)d\omega = n, \tag{2.61}$$

where $f(\omega)d\omega$ is the oscillator strength and n is the full concentration of charge carriers (atoms and electrons). Equation (2.61) is a concise form of Kramers–Kronig relations (see 2.10 and 2.11) and implies that the integral of the dynamic conductivity spectrum (or simply the area under the conductivity curve) taken in the infinite frequency range is conserved and is proportional to the concentration of the charges in a unit volume of matter. Equation (2.61) works for any system (metallic or dielectric), irrespective of the type of interactions, the charge carriers, the temperature, the statistics, or even the presence of a magnetic field, thus providing the most general form of the sum rule [15].

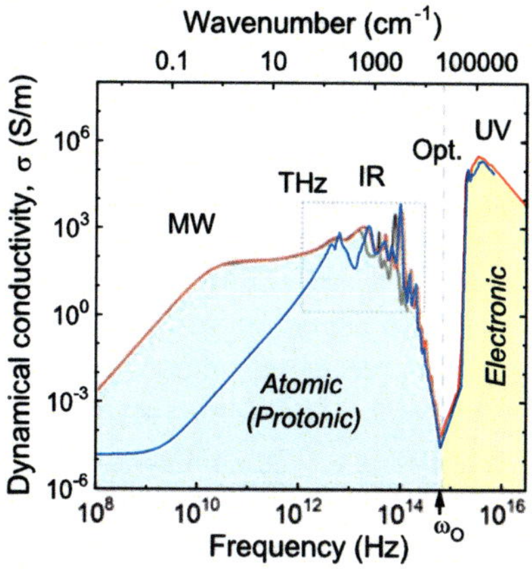

Fig. 2.28 The ultra-broadband spectra of the dynamic conductivity of water (red), ice (blue), and heavy water (gray) on double-logarithmic scales. The dashed vertical line shows the optical cutoff frequency, ω_O, which separates the electronic-conductivity contribution and the protonic-conductivity contribution. The dotted area depicts the infrared region with the highest protonic-conductivity level, which is shown separately in Fig. 2.29

From (2.61), one gets

$$\int_0^{\infty} \sigma(\omega)d\omega = \frac{\pi}{2}\frac{nq^2}{m*} = \frac{\pi}{2}\omega_p^2\varepsilon_0, \tag{2.62}$$

where $\omega_p = \sqrt{nq^2/(m\epsilon_0)}$ is the plasma frequency. Equation (2.62) assumes that the integration of $\sigma(\omega)$ has to be carried over all ranges of frequencies, and that integral of the conductivity spectrum is proportional to the concentration of charge carriers.

Figure 2.28 shows the dynamic conductivity spectra of ice, light water, and heavy water plotted together. All spectra demonstrate a remarkable "transparency window" in the optical region near the frequency $\omega_O \approx 10^{15}$ Hz, which divides the dielectric response into two parts. The right-hand part (UV and X-ray region) has the maximal intensity within the spectrum, two orders of magnitude larger than that for the left-hand part with the maximum intensity in the IR region. The level of conductivity at the "bottom" of the transparency window is ten orders of magnitude lower than that in the UV region, and of the same value as the DC conductivity.[18]

The spectral transparency window is a unique feature of water and a few other substances, such as alcohols and some polymers. It results from the fact that electronic and atomic subsystems do not overlap, as happening in other frequently used

[18]The spectral transparency window allows animals to see under and through the water, and makes water transparent for light of optical frequencies. If our eyes were sensitive to IR frequencies instead of optical, water would be completely opaque. Obviously, evolution chooses the optical region for vision due to the special electrodynamic properties of water.

dielectrics. The spaced-apart electronic and atomic contributions make it possible to apply the rule defined by (2.62) separately for the atomic part only, or, in other words, to apply the equation to a limited frequency range.

As the maximum level of the conductivity of the atomic part is observed in the IR region (see dashed area in Fig. 2.28), the microwave part of the spectrum does not add too much to the integral intensity. Thus, the atomic dynamics in the IR part, which corresponds to the vibrationally averaged dynamics, roughly reduces to the dynamics of protons in the frame of reference of oxygen atoms. Taking into account the fact that the proton mass m_p is an order of magnitude lower than that of oxygen m_O the effective mass m^* is equal to $m_p \cdot m_O/(m_p+m_O) = 1{\cdot}18/(1+18){\cdot}\mathrm{m}_p = 0.95{\cdot}m_p \approx m_p$. Thus, one can temporarily exclude oxygen atoms from the analysis and consider only hydrogen atoms as the primary charge carriers. In other words, (2.62) can be reformulated [71] for the charges in a unit volume of matter. From (2.61), one gets

$$n_p = \frac{2}{\pi}\frac{m_p}{q_p^2}\int\limits_0^{\omega_0} \sigma(\omega)d\omega, \tag{2.63}$$

where n_p and q_p are the concentration and the effective charge of protons, respectively.

Figure 2.29a shows the broadband conductivity spectra of water at different temperatures, supplemented by their partial integrals $S = \int_0^{\omega_{co}} \sigma(\omega)d\omega$, depending on the cutoff frequency, ω_{co}. The high-frequency value of integrals, S_∞, corresponds to the frequency $\omega_{co} = \omega_O$, where ω_O has been defined above (see Fig. 2.28), and is equal to $S \approx 5{\cdot}10^{17}$ S/m·Hz for all forms of water. Using this value, one gets from (2.62) that $n_p \approx 5{\cdot}10^{28}\ \mathrm{m}^{-3}$, which is the concentration of all hydrogen atoms: $n_p^{total} \approx 6{\cdot}10^{28}$ m^{-3} $(2 \cdot 55.5 = 111\ \mathrm{mol/l})$. Thus, the conductivity sum rule, defined by (2.63), works for the protonic subsystem (intramolecular and intermolecular), which is responsible for the dielectric response of water and ice at frequencies below $\omega_O \approx 10^{15}$ Hz.

Figure 2.29b shows the dynamic conductivity spectra $\sigma(\omega)$ of light (H_2O) and heavy (D_2O) water, and their partial integrals S. The spectra differ in the IR part, i.e., the part where the intramolecular dynamics occur, while the relaxation part (below about $500\,\mathrm{cm}^{-1}$) remains unaffected by the isotopic substitution. The main IR peaks of heavy water are shifted by a factor of $a = \sqrt{2} = 1.4$ (see Sect. 2.8 for details). The S_∞ value for D_2O is exactly half that for H_2O, which confirms the validity of (2.62), as the replacement of protons by the twice-as-heavy deuterons affects the S value by a factor of $b = 2$. The S spectrum of D_2O, normalized by coefficients a and b, coincides with that for H_2O (see the dashed line in Fig. 2.29b). Thus, the partial sum rule works separately for protons/deutrons independently on the electronic contribution. In other words, the left-hand side of the water conductivity spectrum reflects the dynamics of protons in the frame of reference of oxygen atoms. Table 2.11 contains the parameters of the sum rule, defined by (2.62), for the spectra shown in Figs. 2.28 and 2.29.

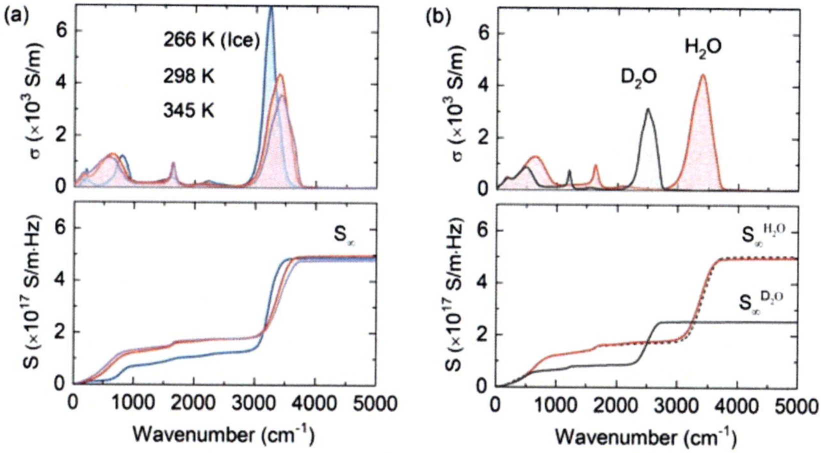

Fig. 2.29 The infrared spectra of the dynamic conductivity, σ, (upper panels) and the corresponding partial integrals S (bottom panels) as a function of the cutoff frequency for: **a** water and ice at different temperatures and **b** light and heavy water at room temperature. S_∞ shows the maximum integral value, which corresponds to the total integral of the spectra. Parameters are in Table 2.11

Table 2.11 Parameters of (2.11) for the conductivity spectra of water, ice, and heavy water at different temperatures

Temperature (K)	H_2O			D_2O
	266 (Ice)	298	345	298
m_p, m_d ($\times$ 10^{-27} kg)	1.67			3.34
ω_O (Hz)	$1.8{\cdot}10^{14}$			
q^* ($\times 10^{-19}$ C)	0.64·1.6=1.02			
S_∞ ($\times$ 10^{17} S/m·Hz)	4.93	5.05	4.84	2.55
n_p / n_d ($\times$ 10^{28} m^{-3})	5.1	5.3	5.2	5.3
n_p / n_d (mol/l)	84.6	87.5	84.6	88.3

The step-like behavior of the S function (see Fig. 2.29) allows one to analyze the contributions of individual IR bands to the integral intensity. Figure 2.30 shows the IR spectra of ice and water. The spectra look different than those in Fig. 2.27 because a logarithm scale is applied to the vertical axis in order to highlight the low-conductivity values. Taking S_∞ as 100%, one can assign the corresponding contributions to the individual peaks. The 3,500-cm^{-1} band gives the main, about 63%, contribution; the libration mode ν_L provides about 22%; and the bending mode ν_2 is about 7%. Other peaks give a few percent altogether. Interestingly, the contribution of intermolecular dynamics, which is shown as the shaded area, constitutes mainly the mode ν_s and contributes a few percent too. This contribution is small, but important for further analysis.

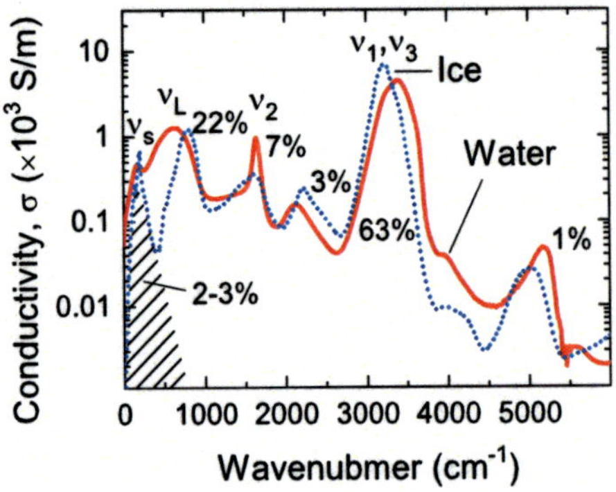

Fig. 2.30 The infrared spectra of dynamic conductivity, σ. Numbers are the contribution of the corresponding mode to the integral intensity, S_{∞}, in percent. The shaded area shows the contribution from the intermolecular dynamics

Thus, part of spectra of ordinary water, heavy water, and ice, below the optical frequencies, satisfies the sum rule: the concentration of protons n_p, which contributes to the spectra, is approximately equal to the total number of protons per unit volume of water/ice, and equal for light and heavy water. The area of optical transparency in the vicinity of 10^{15} Hz separates two global bands of electronic and protonic electromagnetic "activity." The unique dielectric properties of water are, thus, due to the dynamics of protons, intramolecular and intermolecular, in the frame of reference of water molecules. Proton/hole dynamics and the corresponding ionic model of water are considered in Chap. 4.

References

1. P.R. Smith, D.H. Auston, M.C. Nuss, Subpicosecond photoconducting dipole antennas. IEEE J. Quant. Electron. **24**, 255–260 (1988)
2. L.D. Landau, E.M. Lifshitz, *Electrodynamics of Continuous Media* (Pergamon Press, New York, 1958)
3. H. Torii, Time-domain calculations of the polarized Raman spectra, the transient infrared absorption anisotropy, and the extent of delocalization of the OH stretching mode of liquid water. J. Phys. Chem. A **110**, 9469–9477 (2006)
4. O.S. Heavens, *Optical Properties of Thin Solid Films* (Dover, 1955)
5. P. Lunkenheimer, V. Bobnar, A.V. Pronin, A.I. Ritus, A.A. Volkov, A. Loidl, Origin of apparent colossal dielectric constants. Phys. Rev. B. **4**, 052105 (2002)
6. F. Kremer, A. Schönhals (eds.), *Broadband Dielectric Spectroscopy* (Springer, Berlin, 2003)
7. P. Debye, *Polar Molecules* (Chemical Catalog Co., New York, 1929)
8. W.J. Ellison, Permittivity of pure water, at standard atmospheric pressure, over the frequency range 0–25 THz and the temperature range 0–100 °C. J. Phys. Chem. Ref. Data. **36**, 1–18 (2007)
9. H. Yada, M. Nagai, K. Tanaka, Origin of the fast relaxation component of water and heavy water revealed by terahertz time-domain attenuated total reflection spectroscopy. Chem. Phys. Lett. **464**, 166–170 (2008)
10. V.G. Artemov, A unified mechanism for ice and water electrical conductivity from direct current to terahertz. Phys. Chem. Chem. Phys. **21**, 8067–8072 (2019)
11. H. Yada, M. Nagai, K. Tanaka, Determination of the complex dielectric constant of an epithelial cell monolayer in the terahertz region. Chem. Phys. Lett. **464**, 166–170 (2008)

12. J.B. Hasted, *Aqueous Dielectrics* (Chapman and Hall, London, 1973)
13. A. von Hippel, The dielectric relaxation spectra of water, ice, and aqueous solutions, and their interpretation 1. Critical survey of the status-quo for water. IEEE Tnans. Electr. Insul. **23**, 801–816 (1988)
14. K. Okada, M. Yao, Y. Hiejima, H. Kohno, Y. Kajihara, Dielectric relaxation of water and heavy water in the whole fluid phase. J. Chem. Phys. **110**, 3026–3036 (1999)
15. N. Agmon, Tetrahedral displacement: the molecular mechanism behind the Debye relaxation in water. J. Phys. Chem. **100**, 1072–1080 (1996)
16. J.S. Hansen, A. Kisliuk, A.P. Sokolov, C. Gainaru, Identification of structural relaxation in the dielectric response of water. Phys. Rev. Lett. **116**, 237601–5 (2016)
17. R. Kubo, Statistical-Mechanical Theory of Irreversible Processes. I. General Theory and Simple Applications to Magnetic and Conduction Problems. J. Phys. Soc. Jpn. **12**, 570–586 (1957)
18. C. Jaccard, Thermodynamics of irreversible processes applied to ice. Phys. Kond. Mater. **3**, 99–118 (1964)
19. V.G. Artemov, A.A. Volkov, A.V. Pronin, Electrical properties of water: a new insight. Biophysics **59**, 520–523 (2014)
20. V.F. Petrenko, R.W. Whitworth, *Physics of Ice* (University Press, Oxford, 1999)
21. I. Popov, P. Ben Ishai, A. Khamzin, Yu. Feldman, The mechanism of the dielectric relaxation in water. Phys. Chem. Chem. Phys. **18**, 13941–13953 (2016)
22. E.H. Grant, Relationship between relaxation time and viscosity for water. J. Chem. Phys. **26**, 1575–1577 (1957)
23. V.G. Artemov, A. Ryzhov, E. Carlsen, P. Kapralov, H. Ouerdane, Nonrotational mechanism of polarization in alcohols. J. Phys. Chem. B **124**, 11022–11029 (2020)
24. M.P. Kruger, Systematic THz dielectric spectroscopy of ionic liquids. Ph.D thesis 2010
25. A. Pimenov, P. Lunkenheimer, A. Loidl, Broadband dielectric spectroscopy on glycerol and CKN. Ferroelectrics **176**, 33–41 (1996)
26. C.G. Malmberg, A.A. Maryott, Dielectric Constant of Water from 0 to 100 C. J. Res. Nat. Bur. Stand. **56**, 2641–2649 (1956)
27. G.P. Johari, E. Whalley, The dielectric properties of ice Ih in the range 272–133 K. J. Chem. Phys. **75**, 1333–1340 (1981)
28. M. Uematsu, E.U. Franck, Static dielectric constant of water and steam. J. Phys. Chem. Ref. Data. **9**, 1291–1306 (1980)
29. E. Moore, V. Molinero, Structural transformation in supercooled water controls the crystallization rate of ice. Nature **479**, 506–508 (2011)
30. H. Fröhlich, *Theory of Dielectrics* (Oxford University Press, Oxford, 1949)
31. A.K. Soper, The radial distribution functions of water and ice from 220 to 673 K and at pressures up to 400 MPa. Chem. Phys. **258**, 121–137 (2000)
32. J.G. Kirkwood, The local field in dielectrics. Ann. New York Acad. Sci. **40**, 315–320 (1940)
33. C.J.F. Böttcher, *Theory of Electric Polarization* (Elsevier, New York, 1952)
34. L. Onsager, Electric moments of molecules in liquids. J. Am. Chem. Soc. **58**, 1486–1493 (1936)
35. J.G. Kirkwood, The dielectric polarization of polar liquids. J. Chem. Phys. **7**, 911–920 (1939)
36. S.W. Peterson, H.A. Levy, A single-crystal neutron diffraction study of heavy ice. Acta Cryst. **10**, 70–76 (1957)
37. C.G. Hersberg, *Infrared and Rarman Spectra* (D. van Nostrand and Co., New York, 1945)
38. J.L. Aragones, L.G. MacDowell, C. Vega, Dielectric constant of ices and water: a lesson about water interactions. J. Phys. Chem. A. **115**, 5745–5758 (2011)
39. H. Gränicher, Gitterfehlordnung und physikalische Eigenschaften hexagonaler und kubischer Eiskristalle. Z. Kristallogr. **110**, 432–471 (1958)
40. C. Jaccard, Étude théorique et expérimentale des propriétés électriques de la glace. Helv. Phys. Acta. **32**, 89–128 (1959)
41. N. Bjerrum, Structure and properties of ice. Science **115**, 385–390 (1952)

42. V.G. Artemov, E. Uykur, S. Roh, A.V. Pronin, H. Ouerdane, M. Dressel, Revealing excess protons in the infrared spectrum of liquid water. Sci. Rep. **10**, 11320–9 (2020)
43. C. Bai, J. Herzfeld, Special Pairs Are Decisive in the Autoionization and Recombination of Water. J. Phys. Chem. B. **121**, 4213–4219 (2017)
44. F. Perakis, L. De Marco, A. Shalit, F. Tang, Z.R. Kann, T.D. Kühne, R. Torre, M. Bonn, Y. Nagata, Vibrational spectroscopy and dynamics of water. Chem. Rev. **116**, 7590–7607 (2016)
45. C.H. Townes, A.L. Schawlow, *Microwave Spectroscopy* (Dover Publications, New York, 2013)
46. B.M. Auer, J.L. Skinner, Water: hydrogen bonding and vibrational spectroscopy, in the bulk liquid and at the liquid/vapor interface. Chem. Phys. Lett **470**, 13–20 (2009)
47. A.A. Kananenka, J.L. Skinner, Fermi resonance in OH-stretch vibrational spectroscopy of liquid water and the water hexamer. J. Chem. Phys. **148**, 244107–13 (2018)
48. D. Eisenberg, W. Kauzmann, *The Structure and Properties of Water* (Oxford University Press, Oxford, 1969)
49. W.C. Röntgen, Ueber die Constitution des flüssigen Wassers. Ann. Phys. **281**, 91–97 (1892)
50. G.E. Walrafen, Raman spectral studies of the effects of temperature on water structure. J. Chem. Phys. **47**, 114–126 (1967)
51. W.F. Murphy, H.J. Bernstein, Raman spectra and an assignment of the vibrational stretching region of water. J. Phys. Chem. **76**, 1147–1152 (1972)
52. W.B. Monosmith, G.E. Walrafen, Temperature dependence of the Raman OH-stretching overtone from liquid water. J. Chem. Phys. **81**, 669–674 (1984)
53. D.A. Schmidt, K. Miki, Structural correlations in liquid water: a new interpretation of IR spectroscopy. J. Phys. Chem. A. **111**, 10119–10122 (2007)
54. C.S. Choe, J. Lademann, M.E. Darvin, Depth profiles of hydrogen bound water molecule types and their relation to lipid and protein interaction in the human stratum corneum in vivo. Analyst **141**, 6329–6337 (2016)
55. P.L. Geissler, Temperature dependence of inhomogeneous broadening: on the meaning of isosbestic points. J. Am. Chem. Soc. **127**, 14930–14935 (2005)
56. B.M. Auer, J.L. Skinner, IR and Raman spectra of liquid water: theory and interpretation. J. Chem. Phys. **128**, 224511–12 (2008)
57. J.D. Smith, C.D. Cappa, K.R. Wilson, R.C. Cohen, P.L. Geissler, R.J. Saykally, Unified description of temperature-dependent hydrogen-bond rearrangements in liquid water. PNAS **102**, 14171–14174 (2005)
58. M. Falk, T.A. Ford, Infrared spectrum and structure of liquid water. Can. J. Chem. **44**, 1699–1707 (1966)
59. Y. Maréchal, Infrared spectra of water. I. Effect of temperature and of H/D isotopic dilution. J. Chem. Phys. **95**, 5565–5573 (1991)
60. J.J. Max, C. Chapados, Isotope effects in liquid water by infrared spectroscopy. J. Chem. Phys. **116**, 4626–4642 (2002)
61. G.E. Walrafen, Raman spectral studies of water structure. J. Chem. Phys. **40**, 3249–3256 (1964)
62. K. Abe, T. Shigenari, Raman spectra of proton ordered phase XI of ICE I. Translational vibrations below 350 cm^{-1}. J. Chem. Phys. **134**, 104506–11 (2011)
63. J.M. Headrick, E.G. Diken, R.S. Walters, N.I. Hammer, R.A. Christie, J. Cui, E.M. Myshakin, M.A. Duncan, M.A. Johnson, K.D. Jordan, Spectral signatures of hydrated proton vibrations in water clusters. Science **308**, 1765–9 (2005)
64. W. Kulig, N. Agmon, A 'clusters-in-liquid' method for calculating infrared spectra identifies the proton-transfer mode in acidic aqueous solutions. Nat. Chem. **5**, 29–35 (2013)
65. H.R. Zelsmann, Temperature dependence of the optical constants for liquid H_2O and D_2O in the far IR region. J. Mol. Struct. **350**, 95–114 (1995)
66. D.C. Elton, M. Fernandez-Serra, The hydrogen-bond network of water supports propagating optical phonon-like modes. Nat. Commun. **7**, 10193–8 (2016)
67. R. Buchner, J. Barthel, J. Stauber, The dielectric relaxation of water between 0 °C and 35 °C. Chem. Phys. Lett. **306**, 57–63 (1999)

68. M.R. Querry, B. Curnutte, D. Williams, Refractive index of water in the infrared. J. Opt. Soc. Am. **59**, 1299–1305 (1969)
69. N. Agmon, Infrared spectroscopy: the acid test for water structure. Nat. Chem. **8**, 206–207 (2016)
70. D.J. Segelstein, The complex refractive index of water. M.S. Thesis, University of Missouri, Kansas City 1981
71. J.E. Bertie, Z. Lan, Infrared intensities of liquids XX: the intensity of the OH stretching band of liquid water revisited, and the best current values of the optical constants of $H_2O(l)$ at 25 °C between 15,000 and 1 cm^{-1}. Appl. Spectrosc. **5**, 1047–1057 (1996)
72. V.G. Artemov, Dielectric spectrum of water as a proton dynamics response. Bull. Lebedev Phys. Inst. **42**, 187–191 (2015)
73. U. Møller, D.G. Cooke, K. Tanaka, P.U. Jepsen, Terahertz reflection spectroscopy of Debye relaxation in polar liquids. J. Opt. Soc. Am. B. **26**, A113–A125 (2009)
74. K. Shiraga, K. Tanaka, T. Arikawa, S. Saito, Y. Ogawa, Reconsideration of the relaxational and vibrational line shapes of liquid water based on ultrabroadband dielectric spectroscopy. Phys. Chem. Chem. Phys. **20**, 26200–26209 (2018)
75. A.A. Volkov, V.G. Artemov, A.A. Volkov (Jr), N.N. Sysoev, Possible mechanism of molecular motion in liquid water from dielectric spectroscopy data. J. Mol. Liq. **248**, 564–568 (2017)
76. L. Shi, Y. Ni, S.E.P. Drews, J.L. Skinner, Dielectric constant and low-frequency infrared spectra for liquid water and ice Ih within the E3B model. J. Chem. Phys. **141**, 084508–10 (2014)
77. M. Heyden, J. Sun, S. Funkner, G. Mathias, H. Forbert, M. Havenith, D. Marx, Dissecting the THz spectrum of liquid water from first principles via correlations in time and space. Proc. Natl. Acad. Sci. USA **107**, 12068–12073 (2010)
78. S. Carlson, F.N. Brünig, P. Loche, D.J. Bonthuis, R.R. Netz, Exploring the absorption spectrum of simulated water from MHz to infrared. J. Phys. Chem. A. **124**, 5599–5605 (2020)
79. H. Torii, Intermolecular electron density modulations in water and their effects on the far-infrared spectral profiles at 6 THz. J. Phys. Chem. B. **115**, 6636–6643 (2011)
80. G.E. Walrafen, Y.C. Chu, G.J. Piermarini, Low-frequency Raman scattering from water at high pressures and high temperatures. J. Phys. Chem. **100**, 10363–10372 (1996)
81. D. Decka, G. Schwaab, M. Havenith, THz/FTIR fingerprint of the solvated proton: evidence for Eigen structure and Zundel dynamics. Phys. Chem. Chem. Phys. **17**, 11898–11907 (2015)
82. D.F. Coker, R.O. Watts, Structure and vibrational spectroscopy of the water dimer using quantum simulation. J. Phys. Chem. **91**, 2513–2518 (1987)
83. G.E. Walrafen, Raman spectrum of water: transverse and longitudinal acoustic modes below $\approx 300\ cm^{-1}$ and optic modes above $\approx 300\ cm^{-1}$. J. Phys. Chem. **94**, 2237 (1990)
84. D.M. Carey, G.M. Korenowski, Measurement of the Raman spectrum of liquid water. J. Chem. Phys. **108**, 2669–2675 (1998)
85. G.E. Walrafen, Raman spectral studies of the effects of temperature on water and electrolyte solutions. J. Chem. Phys. **44**, 1546–1558 (1966)
86. P. Brüesch, L. Pietronero, S. Strässler, H.R. Zeller, Brownian motion in a polarizable lattice: application to superionic conductors. Phys. Rev. B **15**, 4631–4637 (1977)
87. M.B. Salamon (ed.), *Physics of Superionic Conductors* (Springer, Berlin, 1979)
88. H. Lapid, N. Agmon, M.K. Petersen, G.A. Voth, A bond-order analysis of the mechanism for hydrated proton mobility in liquid water. J. Chem. Phys. **122**, 14506–11 (2005)
89. J. Kim, U.W. Schmitt, J.A. Gruetzmacher, G.A. Voth, N.E. Scherer, The vibrational spectrum of the hydrated proton: comparison of experiment, simulation, and normal mode analysis. J. Chem. Phys. **116**, 737–746 (2002)
90. M. Eigen, Proton transfer, acid-base catalysis, and enzymatic hydrolysis. Part I: elementary processes. Angew. Chem. Int. Ed. Engl. **3**, 1–19 (1964)
91. C. Rønne, P.O. Åstrand, S.R. Keiding, THz spectroscopy of liquid H_2O and D_2O. Phys. Rev. Lett. **82**, 2888–2891 (1999)
92. I. Ohmine, Liquid water dynamics: collective motions, fluctuation, and relaxation. J. Phys. Chem. **99**, 6767 (1995)

93. M. Neumann, The dielectric constant of water. Computer simulations with the MCY potential. J. Chem. Phys. **82**, 5663 (1985)
94. G.E. Walrafen, M.S. Hokmabadi, W.-H. Yang, Raman isosbestic points from liquid water. J. Chem. Phys. **85**, 6964–6969 (1986)
95. M.-C. Bellissent-Funel, Is there a liquid-liquid phase transition in supercooled water? Europhys. Lett. **42**, 161–166 (1998)
96. T. Elsaesser, H.J. Bakker (eds.), *Ultrafast Hydrogen Bonding Dynamics and Proton Transfer Processes in the Condensed Phase* (Kluwer Academic Publishers, NY, 2002)
97. P. Foggi, M. Bellini, D.P. Kien, I. Vercuque, R. Righini, Relaxation dynamics of water and HCl aqueous solutions measured by time-resolved optical Kerr effect. J. Phys. Chem. A. **101**, 7029–7035 (1997)
98. Y.L. Rezus, H.J. Bakker, On the orientational relaxation of HDO in liquid water. J. Chem. Phys. **123**, 114502–7 (2005)
99. A. Arbe, P. Malo de Molina, F. Alvarez, B. Frick, J. Colmenero, Dielectric susceptibility of liquid water: microscopic insights from coherent and incoherent neutron scattering. Phys. Rev. Lett. **117**, 185501–5 (2016)
100. N. Ockman, The infra-red and Raman spectra of ice. Adv. phys. **7**, 199–220 (1958)
101. P.A. Egelstaff (ed.), *Thermal Neutron Scattering* (Academic Press, New York, 1962)
102. R.G. Gordon, Molecular motion in infrared and Raman spectra. J. Chem. Phys. **43**, 1307–1312 (1965)
103. V.I. Gaiduk, B.M. Tseitlin, ChM Briskina, D.S.F. Crothers, Simplified theory of wideband spectra of liquid H_2O and D_2O (from 0 to 1000 cm^{-1}) due to reorienting polar and vibrating H-bonded water molecules. J. Mol. Struct. **606**, 9–27 (2002)
104. E.W. Baumann, Conductivity of D_2O and micromolar DNO_3 in D_2O from 0 to 100 °C. J. Chem. Eng. Data. **38**, 12–15 (1993)
105. Y. Maréchal, Infrared spectra of water. II: dynamics of $H_2O(D_2O)$ molecules. J. de Physique II, EDP Sci. **3**, 557–571 (1993)
106. L. De Marco, W. Carpenter, H. Liu, R. Biswas, J.M. Bowman, A. Tokmakoff, Differences in the vibrational dynamics of H_2O and D_2O: observation of symmetric and antisymmetric stretching vibrations in heavy water. J. Phys. Chem. Lett. **7**, 1769–1774 (2016)
107. J.H. Choi, M. Cho, Computational IR spectroscopy of water: OH stretch frequencies, transition dipoles, and intermolecular vibrational coupling constants. J. Chem. Phys. **138**, 174108–17 (2013)
108. Z. Wang, A. Pakoulev, Y. Pang, D.D. Dlott, Vibrational substructure in the OH stretching transition of water and HOD. J. Phys. Chem. A. **108**, 9054–9063 (2004)
109. A.C. Belch, S.A. Rice, The OH stretching spectrum of liquid water: a random network model interpretation. J. Chem. Phys. **78**, 4817–4823 (1983)
110. J.B. Brubach, A. Mermet, A. Filabozzi, A. Gerschel, P. Roy, Signatures of the hydrogen bonding in the infrared bands of water. J. Chem. Phys. **122**, 184509–7 (2005)
111. J.G. Davis, K.P Gierszal, P. Wang, D. Ben-Amotz, Water structural transformation at molecular hydrophobic interfaces. Nature **491**, 582–585 (2012)

Chapter 3
The Interaction of Electromagnetic Waves with Ice

Abstract The physicochemical distinction between ice and water determines processes of various scales from the adaptation of living organisms to the formation of planets. The electrodynamic properties are not an exclusion: despite the fact that only two types of atoms (oxygen and hydrogen) are involved in the ice and water structures, their dielectric responses are quite different. This chapter focuses on the dielectric properties of ice over a wide frequency range, which are compared to that of water. The similarities between ice and water are discussed within a universal microscopic mechanism of their dielectric response that is suitable for both thermodynamic forms of water. The new approach allows one to model the spectra from DC to the terahertz region on the same footing and provides new insights into our understanding of the molecular dynamics in ice and water.

3.1 Dielectric-Terahertz Spectrum of Ice

Figure 3.1 shows the broadband spectrum of ice in terms of permittivity and dynamic conductivity. The spectrum, which covers frequencies from optics to radiowaves, is compared with that for water, shown by dashed lines and previously discussed in Chap. 2. Both spectra are provided at 0 °C, the temperature at which both coexist in thermodynamic equilibrium, and thus can be compared in the same conditions. A parallel analysis of the spectra reveals some interesting results. First of all, the dielectric relaxation band of ice is shifted by about seven orders of frequency magnitude from the gigahertz to kilohertz region.[1] However, the static dielectric constant, $\epsilon(0)$, in spite of this drastic difference between the relaxation times, stays roughly the same (92 for ice and 88 for water). The amazing similarity of the dielectric constants of water and ice is well known but was considered a coincidence [2], until it was shown that this similarity has in fact a deep physical reason [3]. The parallel analysis of the spectra of ice and water is considered in Sect. 3.4. In addition, ice does not show any secondary relaxations, which has been assigned to molecular reorientation

[1]The difference, corresponding to the shifted dielectric band, between the dielectric losses of ice and water in the microwave region is used, for example, for remote sensing of ice melting [1].

V. Artemov, *The Electrodynamics of Water and Ice*, Springer Series in Chemical Physics 124, https://doi.org/10.1007/978-3-030-72424-5_3

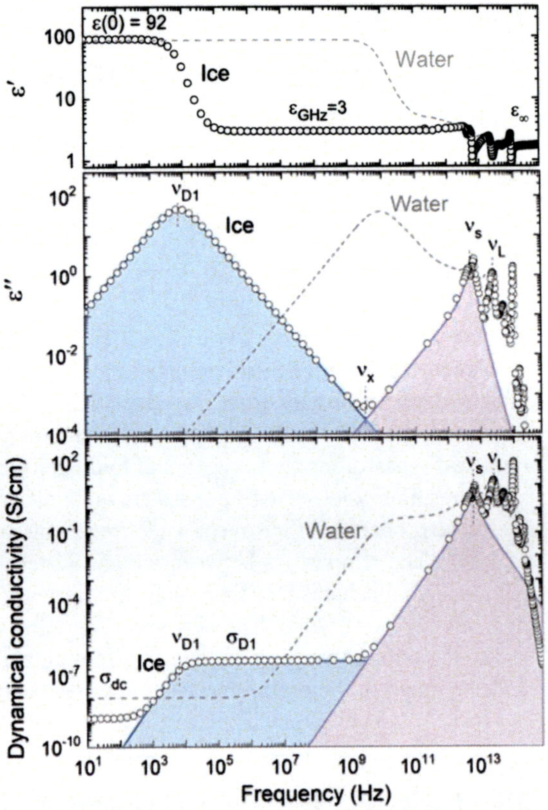

Fig. 3.1 The broadband spectrum of ice near the melting point in terms of real (ϵ') and imaginary (ϵ'') parts of the dielectric permittivity, and dynamic conductivity (σ). The circles represent the experimental data collated from [2–8], the dashed line shows the spectrum of liquid water at the same temperature, and the shaded areas correspond to relaxatory (blue) and oscillatory (magenta) modes with parameters given in Table 3.1. The color areas a spectra components according to (3.1) and (3.2) with parameters given in Table 3.1

following the proton-transfer reactions (see Sect. 2.6.2). The 5 THz mode, which in ice splits into two components, transforms into the main dielectric relaxation as the frequency decreases, and no excess wings are observed.

The dielectric response of ice shown in Fig. 3.1 is well described up to about 30 THz by the following formulas:

$$\varepsilon'(\nu) + i\varepsilon''(\nu) = \varepsilon_\infty + \frac{\Delta\varepsilon_{D1}\nu_{D1}}{\nu_{D1} + i\nu} + \sum_{j=s,L} \frac{\Delta\varepsilon_j \nu_j^2}{(\nu_j^2 - \nu^2) + i\nu\gamma_j}, \tag{3.1}$$

and

$$\sigma(\nu) = \sigma_{dc} + \frac{\sigma_{D1}\nu^2}{\nu_{D1}^2 + \nu^2} + \sum_{j=s,L} \frac{\sigma_j (\nu\gamma_j)^2}{(\nu_j^2 - \nu^2)^2 + (\nu\gamma_j)^2}, \tag{3.2}$$

where the first terms are the high-frequency permittivity and static conductivity, the second terms are Debye relaxation function, and the sums (last terms) are two

Table 3.1 The parameters of the water and ice spectra shown in Fig. 3.1, and decomposed in accordance with (3.1) and (3.2)

	σ_{dc} (μS/m)	σ_{D1} (S/m)	$\Delta\epsilon_{D1}$	ν_{D1} (Hz)	ν_x (GHz)	ν_s (THz)	γ_s (THz)	$\Delta\epsilon_s$	ν_L (THz)	γ_L (THz)	$\Delta\epsilon_L$
Ice (273 K)	0.16	$4.1 \cdot 10^{-5}$	87.1	$7.8 \cdot 10^3$	1.6	6.1	3.30	1.09	23.2	3.09	0.34
Water (273 K)	1.2	40.3	83.4	$8.5 \cdot 10^9$	1800	5.9	4.61	1.35	17.9	13.2	0.80

Lorentz oscillators. Parameter σ_{D1} is defined by 2.38, and σ_j is defined by analogy. Best fit parameters are given in Table 3.1, which compares the results of spectra decomposition for ice and water near the melting point. Frequency ν_x, which for ice equals to 1.6 GHz, corresponds to the minimum of dielectric losses, or the point of the highest transparency.

While the low-frequency parts of the spectra of ice and water undergo a global transformation, the infrared region above 3 THz does not change significantly at phase transition. Figure 3.2 shows the IR spectra of ice and water in (a) normal and (b) logarithmic scales. The spectrum of ice qualitatively repeats that for water with small changes in the positions and widths of the peaks. Note that the maximal dynamic conductivity, $\sigma_{max} \sim 10^2$ S/cm, is observed near 3500 cm^{-1} (10^{14} Hz or 3 μm), which is two orders of magnitude larger that the conductivity in the microwave region. Four fundamental intramolecular modes of water are also present in ice (see Fig. 3.2a). The H–O–H bending mode, ν_2, of ice is located near 6 μm; the combined symmetric and asymmetric OH-stretching mode, $\nu_{1,3}$, lies near 3 μm; and the comparatively weak bands attributed to oscillations with contributions from overlapping overtones and combinations of the fundamental modes lie near 1.5- and 2-μm, respectively [7].

However, there are some minor differences between the IR spectra of liquid and solid water. As one can see in Fig. 3.2b, where the spectrum is plotted in linear scales, the ice mode $\nu_{1,3}$ is shifted to a lower frequency by about 15% in comparison with that for water, and it has a higher amplitude and asymmetric shape. Although it is obvious that this mode is caused by the relative displacement of the proton and the oxygen atom of the same molecule, the analysis of the fine structure of $\nu_{1,3}$ mode is still far from complete. Quantum effects, such as proton tunneling and Fermi resonance (see Sect. 2.7.2), and the partial coupling of the vibrational and phonon modes [9], which manifest themselves differently in water and ice, do not allow us to explicitly interpret the band shape. Nevertheless, one can conclude that the spring constant for O–H stretching is slightly weaker in ice than in water, caused by density difference, and presumably the lower coordination number. On the contrary, the H–O–H bending mode ν_2 is shifted to higher frequencies by several percent, indicating a stiffening of the spring constant of ice, and a higher quality factor of the corresponding oscillation. The libration (hindered rotation) band, which appears at 12 μm (23 THz) is also slightly blueshifted in ice, indicating the higher cooperativity of H_2O molecules in ice than in water, caused presumably by the long-range order formation.

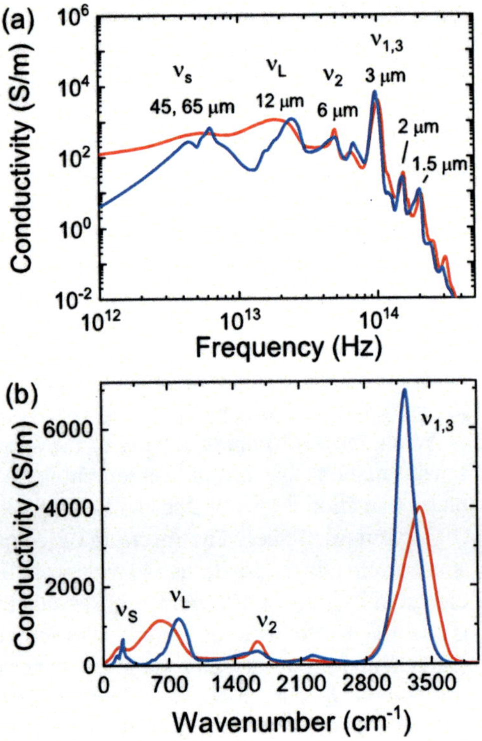

Fig. 3.2 The infrared spectra of ice (blue) and water (red) in two different representations: **a** logarithmic, and **b** linear. Numbers with μm are the wavelengths, which correspond to the frequency of the peak maxima

The mode ν_s near 200 cm^{-1} (5.3 THz), which is associated with the intermolecular proton transfer following the shortening of intermolecular distances (see Sect. 2.6.1), splits into two components for ice that appear at 140 and 190 cm^{-1} (65 and 45 μm), and presumably indicate transverse-longitudinal optical (LO–TO) splitting, which assumes long-ranged Coulomb interactions. Note that the 140 cm^{-1} mode disappears in the Raman spectrum of ice [10], which indirectly confirms this idea. Note that Walrafen concluded [11] that the single Raman peak is connected with longitudinal spherical acoustic dilatational waves and involves the displacement of the molecular center of mass. He identified that optic and multiphononic modes, which mainly involve proton motion, appear above ≈ 330 cm^{-1}, while pure acoustic modes, which involve H_2O center-of-mass motion, appear below ≈ 330 cm^{-1}. Nevertheless, Abe and Shigenari [12] present the first experimental confirmation of the LO–TO splitting of the mode near 200 cm^{-1} in ice. They showed that the splitting of the mode ν_s is 6.0 cm^{-1}, which is close to the value previously reported by Bertie et al. [13], confirming long-range electrostatic interaction in ice. Bertie also showed that the transverse-acoustic (TA) branch appears only as a negligibly small shoulder at 65 cm^{-1} in the direct infrared spectrum of ice, and thus cannot be responsible for the ν_s mode.

Whalley and Klug [14] argue that the strong infrared absorption in the O–H stretching region is already an indicator of LO–TO splitting, which occurs in the vibrational spectrum of ice, as well as in liquid water, due to the unconditional similarity of the spectra. This similarity hints that liquid water should also have collective vibrations that propagate in a similar way to those in ice and that are caused by long-range Coulomb interactions. Thus, the IR spectrum of ice and water can be fully understood only considering the phononic modes, which are usually missing in conventional molecular-dynamic simulations. Note that at terahertz frequencies water behaves more like a moderately rigid, isotropic, elastic solid as highlighted by Walrafen [11], while ice looks more like a crystal.

3.2 The Temperature Dependence of Spectral Parameters

Figure 3.3 shows the temperature dependencies of the main microscopic transport parameters of water and ice: the static conductivity, σ_{dc}, the high-frequency conductivity, σ_{D1}, the relaxation frequency, ν_{D1}, and the self-diffusion coefficient, D_{self}. Curves are collated on the basis of data from [2, 5, 15–18] and represented as Arrhenius plots.[2] The perfect linear behavior of all quantities indicates that in the given temperature intervals, i.e., 0–100 °C for water and $-60 - 0$ °C for ice, both obey Arrhenius law: $A = A_0\exp(E_a/k_BT)$, where A is the transport parameter, A_0 is the pre-exponential factor, and E_a is the activation energy. The corresponding best-fit parameters are given in Table 3.2. The figure and table show that parameters D_{self}, ν_{D1} and σ_{D1} have approximately the same activation energy $E_a \approx 0.16$ eV for water, and $E_a \approx 0.6$ eV for ice. Additionally, at the phase transition, all quantities change synchronously by about 6–7 orders of magnitude, following the spectral shift between ice and water, except the static conductivity σ_{dc}, the activation energy of which is 1.2 and 2.5 times higher than that for other quantities for ice and water, respectively. Interestingly, σ_{dc} changes by one order of magnitude only, and in the megahertz frequency region the conductivity of ice exceeds the conductivity of water by a factor of 35 (see Fig. 3.1). Thus, ice is an order of magnitude more conductive in the megahertz region than water.

Figure 3.4 compares the temperature dependence of H_2O molecule mobility, $\mu_{H_2O} = D_{self}/k_BT$, with the low frequency, σ_{dc}, and the high-frequency, σ_{D1}, conductivities, which are plotted in dimensionless units. One can see that μ_{H_2O}(T) coincides with σ_{D1}(T) but differs from σ_{dc}(T). This means that the high-frequency conductivity mechanism is diffusion controlled, or, in other words, the energy barrier of conduction is the same as for diffusion. Although the static electrical conductivity σ_{dc} is also expected to be diffusion controlled, the higher activation energy than that for H_2O mobility indicates an interaction between charge carriers (H_3O^+ and OH^- ions). This fact is missing in the account of the concentration of hydrogen ions in the concept of pH (see Sect. 1.3.3).

[2]The plot displays the logarithm of a quantity plotted against reciprocal temperature.

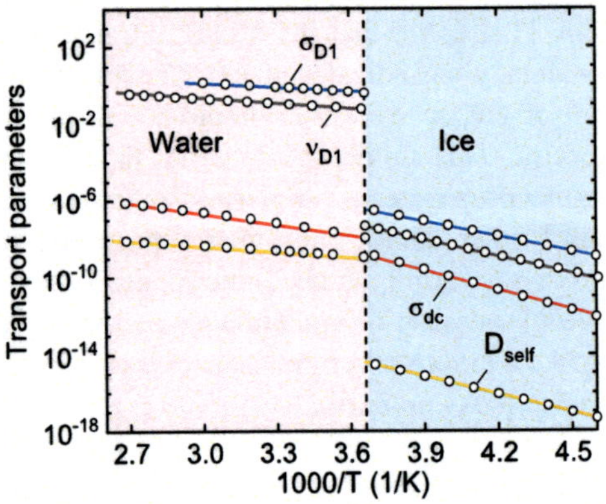

Fig. 3.3 The temperature dependencies of different transport parameters: self-diffusion coefficient, D (m^2/s); static conductivity, σ_{dc} (S/cm), high-frequency conductivity, σ_{D1} (S/cm), and relaxation frequency, ν_{D1}, for water and ice. The dashed vertical line corresponds to the melting point. Circles are experimental data, lines are linear fit with parameters given in Table 3.2

Table 3.2 The parameters of the water and ice spectra shown in Fig. 3.1, and decomposed in accordance with (3.1) and (3.2)

		σ_{dc} (S/m)	σ_{D1} (S/m)	ν_{D1} (THz)	D_{self} (m^2/s)	μ_{H_2O} ($m^2/(V{\cdot}s)$)
Water	A_0	7.90	$4.2 \cdot 10^4$	28.6	$2.3 \cdot 10^{-6}$	$1.9 \cdot 10^{10}$
	E_a (eV)	−0.37	−0.14	−0.19	−0.18	−0.15
Ice	A_0	$1.6 \cdot 10^6$	$1.0 \cdot 10^6$	100	$1.1 \cdot 10^{-3}$	$1.3 \cdot 10^{12}$
	E_a (eV)	−0.70	−0.53	−0.56	−0.61	−0.58

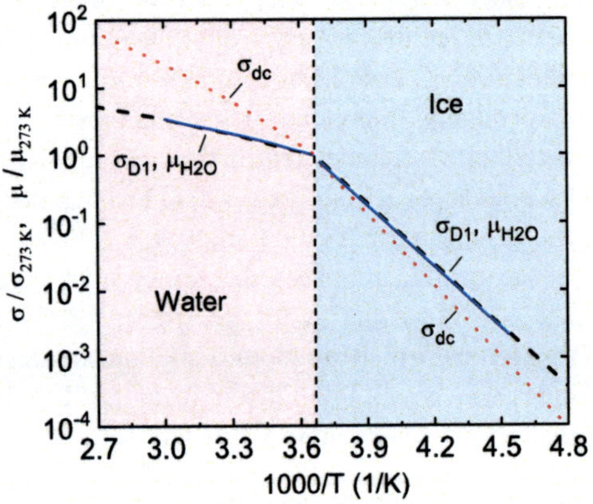

Fig. 3.4 The temperature dependencies of dimensionless conductivity and molecular mobility, for water and ice normalized at the melting point (dashed vertical line)

Different activation energies for high- and low-frequency conductivity in ice forced Bjerrum to conclude [19] that they are caused by different mechanisms. He suggested that σ_{dc} results from the migration of ionic defects, while the high-frequency conductivity, σ_{D1}, results from the migration of H–O· · · O–H, and O–H· · · H–O defects (Bjerrum valence defects, or L-D defects, see Fig. 1.5). The latter are expected to be negatively or positively charged species. The idea of two types of defects producing two plateaus of conductivity is still very popular in the literature [2, 20–23].

However, apart the fact that no direct experimental confirmation of the existence of Bjerrum defects has been found, there are some arguments that hinder their application to the dielectric relaxation phenomena:

- Electrostatic repulsion between the partial positive charges on the H atoms of the D defect and between partial negative charges of the L defect is bound to alter the defect structure [23], which prevents them from contributing to dielectric losses on the timescale of dielectric relaxation.
- The mobility of neutral water molecules is found to be an order of magnitude higher than that expected for L-D defects [24], assuming different mechanisms for self-diffusion and defect diffusion, which is not the case as both have the same activation energy (see Fig. 3.4).
- The concentration of Bjerrum defects in ice is expected to be several orders of magnitude larger than that for ionic defects [2]. In water, this concentration is expected to be even higher [20], close to the full concentration of H_2O molecules. Apart from the fact that such a high concentration of defects calls into question the crystalline structure of ice, it also assumes a strong electrostatic interaction between charge carriers (both ionic and L-D). As ionic species and L-D defects are expected to have different mobilities, their mutual interaction should lead to two frequency-separated dielectric relaxations, which are not observed.
- The expected concentration n_{LD} of Bjerrum defects is 10^{16} cm^3 [2] for the dielectric constant $\epsilon(0) = \epsilon_\infty + 4\pi p^2 n/(3k_B T)$, where p is the dipole moment of the effective dipoles, giving a value that is 6–7 orders of magnitude lower than the experimental one. Thus, the L-D defects alone cannot explain the dielectric relaxation, neither for water nor for ice.

These arguments, among others, do not allow us to rely with confidence on the concept of defects when explaining dielectric phenomena in either water or ice.

An alternative approach, which satisfies the electrodynamic data and does not contradict the basic physicochemical properties of ice, is to consider the two plateaus of frequency-dependent electrical conductivity on the same footing. Such a model was suggested in [25], and further elaborated in [3, 26]. It has been shown that the static conductivity σ_{dc} and the high-frequency conductivity σ_{D1} are both due to the dynamics of H_3O^+ and OH^- ions, but averaged over different periods of time. The model is based on the idea that the mobility of ionic species is time-dependent, and that the two plateaus of conductivity are limiting cases of diffusion with and without mutual interaction. Thus, the dielectric dispersion area between the two plateaus of conductivity (Debye relaxation) corresponds to the transition between these two regimes.

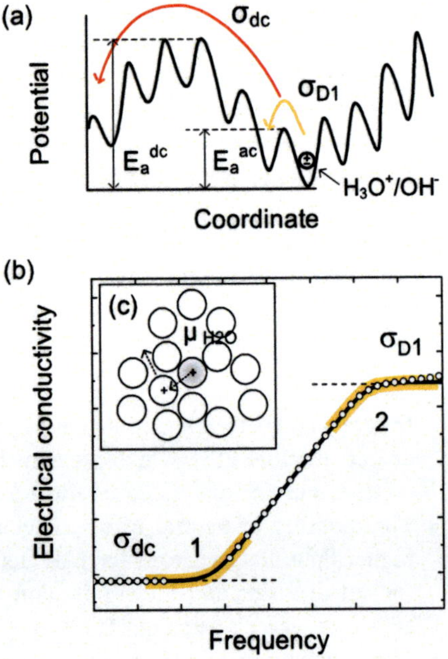

Fig. 3.5 The mechanism of proton conduction in water and ice. **a** Potential for the intrinsic ionic species in ice or water. **b** The corresponding conductivity spectrum. The high-frequency conductivity, σ_{D1}, corresponds to individual ion hopping, while the low-frequency conductivity, σ_{dc}, is due to long-range ion rearrangement. **c** The interstitial mechanism of diffusion of the ionic species

Figure 3.5a shows the effective potential that is used in the model and that is formed by the surrounding molecules of the ionic species (H_3O^+ or OH^-). The potential has two spatial periods, one is due to the nearest molecular species (see Fig. 3.5c), and the other is because of the electrostatic interaction with ions of the opposite charge. The latter was previously neglected in the interpretation of the static conductivity of pure water (see Sect. 1.3). Two characteristic activation energies of diffusion, $E_a^{ac} = 0.14$ eV (see Table 3.2), and $E_a^{dc} = 0.37$ eV are the amplitudes of the potential, which define the high-frequency, σ_{D1}, and the low frequency, σ_{dc}, conductivity plateaus, respectively, depending on the sounding frequency, as shown in Fig. 3.5b.

Within the model, ions obey Brownian motion over short periods of observation,[3] and also at very long (more than 1 μs) periods of observation (the static regime). However, the diffusion coefficients (or mobilities) are different for these two cases. The short-term diffusion limit (high frequencies) is for such short periods that the mutual interaction between ions can be neglected. The long-term diffusion limit (very low frequencies) is, on the contrary, for the ambipolar diffusion of interacting ions. The first case corresponds to the high-frequency plateau, σ_{D1}, while the second leads to the static conductivity plateau σ_{dc}. The intermediate region of dispersion is a transition between these two cases.

[3]The observation time here corresponds to the period of an applied alternating electric field.

Note that the increase of conductivity with the increase of the sounding frequency (region 1 in Fig. 3.5b) is, in fact, a Debye–Falkenhagen effect [27].[4] From the other side, the decrease of the dynamic conductivity as the frequency decreases (region 2 in Fig. 3.5b) is an electrophoretic effect [28].[5]

Thus, this phenomenological model allows the frequency dispersion of electrical conductivity to be interpreted using only one type of charge carrier, namely H_3O^+ and OH^- ions, the mobility of which is time/frequency dependent. The model is suitable for both ice and water. The phenomenology, described above, qualitatively explains the experimentally observed coincidence of the activation energies of molecular mobility, and high-frequency conductivity, σ_{D1}, and allows one to obtain the experimental dielectric constant of ice and water on the same footing [3]. The details of the model are discussed in Sect. 3.5, and the quantitative analysis is given in Chap. 4.

3.3 Ice Among Other Dielectrics

Quartz served as a prototype for the Bernal–Fowler structure of water based on the similarity of their X-ray patterns (see Chap. 1). Here, we compare ice, as well as water, with quartz and other dielectrics from the viewpoint of their electrodynamic properties.

Table 3.3 shows direct-current conductivity σ_{dc}, and the static dielectric constant $\epsilon(0)$, of different ionic and molecular dielectrics, including water and ice. Values are given at room temperature (where possible), or near the melting/transition point. For simple dielectrics, ice and water exhibit relatively high static proton conductivity, which is $\sigma_{dc} \approx 10^{-8}$ S/m for ice, and three orders of magnitude larger for water. No other dielectrics, except ionic liquids and superionic conductors, show such a high ionic conductivity. This fact shows the ability of protons to "tunnel" between water molecules, while the actual mobility of H_3O^+ and OH^- ions as a whole is relatively low. Note that the role of these excess protons (and the corresponding proton holes) is underestimated in the polarization models of water, which seem to combine the properties of molecular and ionic systems.

From an electrodynamic point of view, ice looks similar to AgI, a superionic conductor, which shows high mobility of Ag^+ ions in the lattice formed by iodide (I^-) ions. As shown in Sect. 2.6.1, AgI has also similar Raman spectrum to water and is better suited as a model system for the electrodynamics of ice and water than quartz (SiO_2), which as one can see from Table 3.3 has a relatively low dielectric constant and DC conductivity.

The static dielectric constants of ice and water are also anomalously high, $\epsilon(0) \approx 90$. The value of $\epsilon(0)$ is commonly associated with the ability of water to dilute

[4]The increase of the conductivity when the applied voltage has a very high frequency.

[5]The effect in which the mobility of ions moving under the influence of an applied electric field is affected by the flow of ions of the opposite charge in the opposite direction.

Table 3.3 The electrical parameters of selected dielectrics: dielectric constants, $\epsilon(0)$, DC conductivity, σ_{dc}, and molecular dipole moment, μ_0

Dielectric	State	Temp. (K)	σ_{dc} (S/cm)	$\epsilon(0)$	μ_0 (D)	References
NaCl	Crystal	298	$3.2 \cdot 10^{-20}$	5.9	–	[29]
KBr	Crystal	298	$0.2 \cdot 10^{-20}$	4.9	–	[29]
Ice	Crystal	273	$1.6 \cdot 10^{-11}$	92	1.85	[3]
Water	Liquid	298	$5.5 \cdot 10^{-8}$	78	1.85	[25]
SiO_2	Crystal	298	$\approx 10^{-17}$	3.9	–	[35]
Diamond	Crystal	298	$\approx 10^{-20}$	5.5	–	[35]
$ZrO_2 \cdot Y_2O_3$	crystal	298	$<10^{-9}$	20–35	–	[34]
α-AgI	Crystal	423	1.2	180	–	[31, 32]
β-AgI	Crystal	298	$5.9 \cdot 10^{-6}$	14	-	[31, 32]
$C_2NH_8NO_3$ (EAN)	Liquid	298	$2 \cdot 10^{-2}$	26.3	–	[30]
$(CH_3)_2SO$ (DMSO)	Liquid	298	$3 \cdot 10^{-8}$	46.7	3.9	[33]
C_2H_5OH (Ethanol)	Liquid	298	$37 \cdot 10^{-6}$	26	1.68	[26]

solutes. However, contrary to the commonly accepted opinion, the static dielectric constant is not as strongly correlated with the dipole moment of a molecule as it would be in case of pure rotational polarization (see Fig. 3.6). As shown in [36], only such materials as covalent solids and ionic substances with inversion symmetry obey the Clausius–Mossotti equation:

$$\frac{\varepsilon - 1}{\varepsilon + 2} = \frac{1}{3\epsilon_0} \frac{\alpha_m}{V}, \tag{3.3}$$

where α_m/V is the polarizability per unit volume, which in the case of molecular dielectrics is associated with the molecular dipole moment, and in the case of ionic dielectrics is connected with the relative ion displacement.

However, (3.3) fails to reproduce the temperature dependence of the dielectric constant of molecular systems (including ice and water), or that of many strongly polarizable materials with large values of ϵ, such as ferroelectrics and piezoelectrics. Instead, their dielectric constants obey the Curie–Weiss law:

$$\varepsilon = \frac{C}{T - T_c}, \tag{3.4}$$

where C is the Curie constant and T_c is the Curie temperature.

The temperature dependence of the dielectric constant is easily obtained from (3.4) by differentiation with respect to T at constant pressure P:

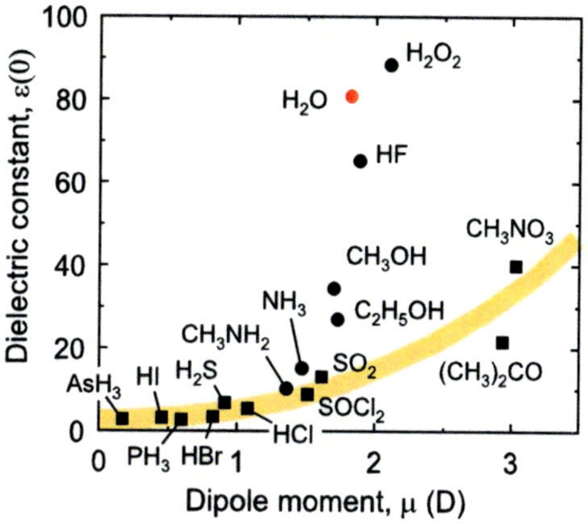

Fig. 3.6 The dielectric constants of liquids as a function of their dipole moments. Squares are for unassociated liquids, and circles are for associated liquids (including water) with high values of the dielectric constant out of the general trend (the orange line). Data from [28]

$$\frac{1}{\varepsilon^2}\left(\frac{\partial \varepsilon}{\partial T}\right)_P = -\frac{1}{C}, \tag{3.5}$$

where the right part is represented by a constant.

As discussed in Chap. 2 (see Figs. 2.9 and 2.10), (3.4) and (3.5) are perfectly suited for both ice and water. From this point of view, ice and water behave more like ferroelectrics, antiferroelectrics, piezoelectrics, and some other crystals that, following (3.5), undergo phase changes, and usually show very large and abrupt variations of the dielectric constant as a function of temperature [37]. The change of ϵ at the transition point depends strongly on the nature of the phase transition. Impurities, defects, and domain boundaries greatly influence the measurement of large dielectric constants [38].

The high value of $\epsilon(0)$ has been shown [36] to be connected with the high value of ϵ_∞ (see Fig. 3.1), which is observed at high frequencies, and caused, in particular, by the high terahertz absorption. Figure 3.7 compares the dynamic conductivity spectra of ice and water with ionic solids and ionic liquids, respectively. The 5 THz oscillatory mode of water and ice, discussed in Sect. 2.6.2, is very similar to those observed in ionic systems at 1–10 THz, which is assigned to the ionic oscillation modes in the case of counter ions. The intensity of this mode in water and ice is of the same order of magnitude as for ionic systems. All spectra show similar shapes with the oscillation and diffusional parts above and below ~1 THz, respectively. Thus, according to the sum rule (see Sect. 2.8) the concentration of the oscillating charges in water is expected to be comparable with that in ionic liquids.

The diffusion part of the ionic conductivity below 1 THz in Fig. 3.7 corresponds to the Debye relaxation modes for water and ice (see Fig. 3.1). The frequency range of this part significantly changes from one material to another, as ionic mobility depends on mass and temperature and essentially varies between liquids and solids. Interestingly, ionic crystals do not show any dipole relaxation effects up to a frequency

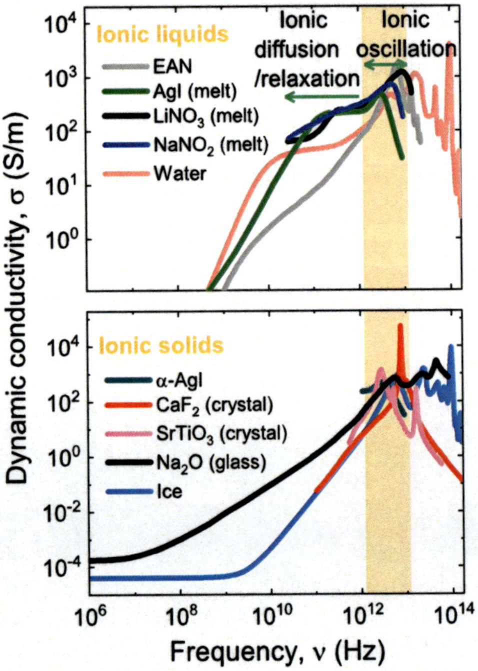

Fig. 3.7 The spectra of ionic dielectrics in comparison with water (red) and ice (blue): liquids and melts (top) and crystals and glasses (bottom). For all the materials shown, the level of alternating current conductivity is of the same order of magnitude as for ice and water. Adapter from [3] with permission from the PCCP Owner Societies

of about 1 GHz. The oscillation modes above 10 THz for ice and water, on the contrary, coincide and have an intramolecular nature. Therefore, the ice and water spectra reflect both molecular and ionic dynamic components.

3.4 Similarities Between Water and Ice

At 0 °C water and ice can coexist, which makes possible to compare their electric properties and search for the microscopic mechanisms responsible for these properties.

Figure 3.8 shows the broadband dielectric spectra of ice and water compiled at 273 K in terms of dielectric permittivity $\epsilon'(\nu)$ and dynamic conductivity $\sigma(\nu)$. The spectra are the same as shown in Fig. 3.1, but presented in a more convenient form for a parallel analysis. The difference between the spectra is obvious. The DC conductivity σ_{dc} and its high-frequency limit σ_{D1} differ by about seven orders of magnitude for water and two orders of magnitude for ice. The region of the dielectric dispersion is shifted from the gigahertz region for water to the kilohertz region for ice. Nevertheless, spectra also have much in common. Both spectra can be fitted using the following model of two oscillators, one of which is overdamped:

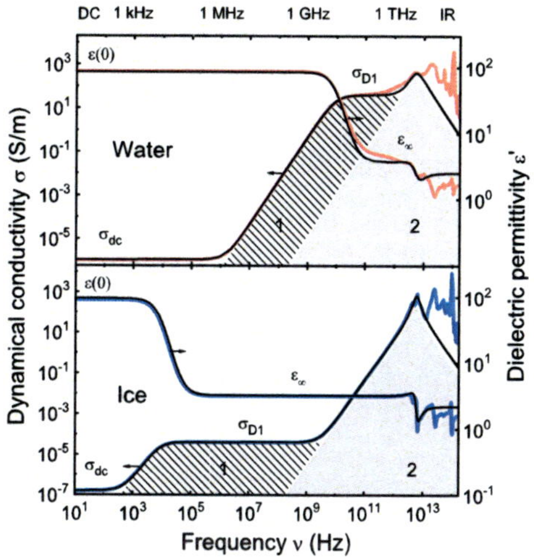

Fig. 3.8 The broadband dielectric spectra of dielectric permittivity $\epsilon'(\nu)$ and the dynamic conductivity $\sigma(\nu)$ of water (top red) and ice (bottom blue) at 273 K. Colored lines are experimental data, black lines are model according to (3.15), (3.16). The shaded areas 1 and 2 are guides for the eyes, plotted according to (3.6) with parameters given in Table 3.5. Adapter from [3] with permission from the PCCP Owner Societies

$$\varepsilon^*(\omega) = i\frac{\sigma^*(\omega)}{\varepsilon_0\omega} = \varepsilon_\infty + \sum_{j=1}^{2} \frac{\Delta\varepsilon_j\omega_j^2}{\omega_j^2 - \omega^2 + i\omega\gamma_j}, \tag{3.6}$$

where $\epsilon^* = \epsilon' + i\epsilon'' = \epsilon' + i(\sigma')/\omega\epsilon_0$, $\omega = 2\pi\nu$ is the angular frequency, ω_j and γ_j are eigenfrequencies and damping constants, and $\Delta\epsilon_j$ is the dielectric contribution of the corresponding mode. The two modes of (3.6) are shown in Fig. 3.7 as shaded areas labeled as "1" and "2." The best-fit parameters are given in Table 3.4. Note that our analysis here does not account for intramolecular infrared vibrations. For simplicity, we also intentionally skip the so-called "second relaxation" of water (see Sect. 2.6.2), which is not found in ice. However, the parameters of the second relaxation are provided in Table 3.4 in brackets and marked as (1′). In fact, we use only one of the possible ways of the spectra decomposition (see Sect. 2.3). Oscillators are not genuine features of the spectrum, but effective tools for analysis, allowing us to compare the spectra of ice and water. Note also that the overdamped harmonic oscillator reduces to relaxation at the low-frequency limit, but the oscillator's central frequency, ν_j, shown in Table 3.4, is not equal to the relaxation frequency, ν_{D1}, shown in Table 2.2.

Table 3.4 shows that mode 1 is a main contributor to the static dielectric constant $\epsilon(0)$ for both ice and water. It provides up to 95% of its value, leaving only a few percent to other spectral features. For instance, mode 2, although it looks large in the conductivity spectrum (see Fig. 3.8), is responsible for only 2% of the static dielectric constant. Nevertheless, this latter mode is very important for understanding the

Table 3.4 The fit parameters of the spectra from Fig. 3.8 according to (3.6). Additionally, $\epsilon(0) = 95$ and 88; $\epsilon_\infty = 2.2$ and 2.5; $\sigma_{dc} = 0.16$ and 1.2 μS/m for ice and water, respectively,

	j	$\Delta\epsilon_j$	ν_j (THz)	γ_j (THz)	σ_j (S/m)
Ice	1	92.2	0.6	$2.4{\cdot}10^6$	$4.1{\cdot}10^{-5}$
	2	1.12	5.7	2.9	628
Water	1	83.4	0.09	0.82	40
	(1′)	(2.1)	(1.7)	(4.5)	(76)
	2	1.15	5.0	4.3	372

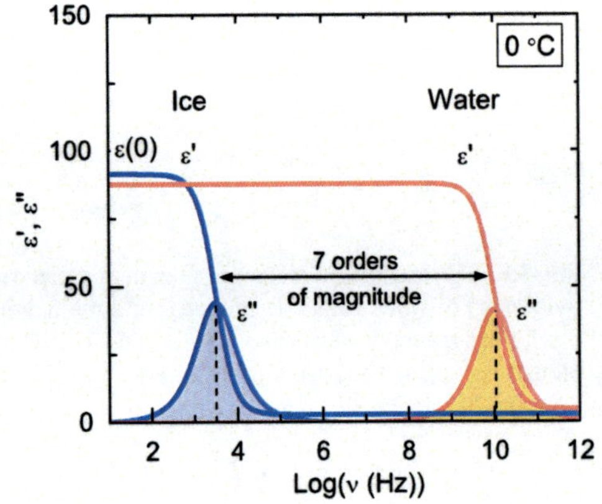

Fig. 3.9 The dielectric spectra of water (red) and ice (blue) at 0 °C in terms of the real, ϵ', and imaginary, ϵ'', parts of the dielectric constant. The spectra look similar with a relative shift of seven orders of the frequency magnitude

microscopic dynamics. Figure 3.8 shows that it remains stable at the phase transition. In particular, the central frequency, width, and area stay roughly the same (see Table 3.4). In contrast, mode 1 behaves differently. Although the total area of mode 1 conserves when water undergoes a phase transition, its central frequency changes dramatically. The equivalent total area of mode 1 for ice and water explains their experimentally observed close dielectric constants: 95 and 88 at 0 °C, respectively. However, the identity of the dielectric constants is puzzling, because, as discussed above (see Sect. 2.3), the microscopic mechanisms of $\epsilon(0)$ in water and ice are historically assumed to be microscopically different. The dielectric constant $\epsilon(0)$ of water is assumed to be due to the reorientations of the molecular H_2O dipoles, while the $\epsilon(0)$ of ice is better understood in terms of the migration of defects [5, 39].

The dramatic similarity of the ice and water dielectric spectra becomes obvious from a comparison of their dielectric responses in terms of the real, ϵ', and imaginary, ϵ'', parts of the dielectric constant, which are shown in Fig. 3.9. Despite the shift of the main relaxation band by seven orders of the frequency magnitude, both spectra look identical. The equivalent areas of the dielectric relaxations result in similar dielectric constants. From an electrodynamic point of view, ice behaves like slowed-down water, the microscopic dynamics of which are similar but delayed.

In addition, the fact that both ice and water have very close dielectric constants, and their spectra have similar structures, shows that they can be analyzed on the same

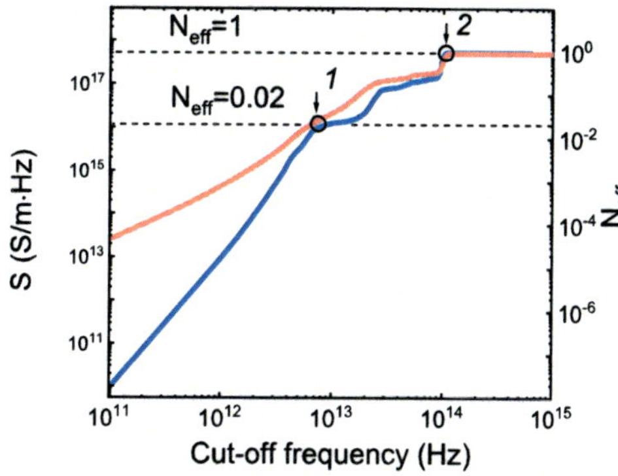

Fig. 3.10 The spectral weight (left scale) and the effective number of conducting particles per the volume of one H_2O molecule (right scale) according to (3.8) versus the cutoff frequency ω_0. Adapter from [3] with permission from the PCCP Owner Societies

footing [3]. As discussed in Sect. 2.8, the general property of AC conductivity is the sum rule, which states that the integral of the conductivity spectrum is a constant quantity that depends on the concentration of the charge carriers and effective charge only. The atomic-molecular dynamics of both ice and water are separated from the electronic contribution by the optical "transparency window" (see Fig. 2.28). This allows one to separate the electronic and atomic contributions to the conductivity spectrum.

Figure 3.10 shows the low-energy spectral weight S:

$$S = \int_0^{\omega_0} \sigma(\omega')d\omega' = \frac{\pi}{2} \sum_i \frac{n_i q_i^2}{m_i}, \tag{3.7}$$

which is plotted as a function of the cutoff frequency ω_0. Note that the integral defined by (3.7) depends only on the charged-particle-related parameters. When ω reaches $\omega_0 \approx 10^{14}$ Hz, S stabilizes at the level of $S_\infty \approx 5.5 \cdot 10^{17}$ S/m·Hz, which coincide for ice and water. Substituting proton mass $m_p = 1.67 \cdot 10^{-27}$ kg and the effective charge $q^* = 1.0 \cdot 10^{-19}$ C (about 60% of the charge of the proton [40]), we obtain the concentration of carriers for ice and water: $n = 111$ mol/l or $6 \cdot 10^{28}$ m^{-3}. This concentration equals the total concentration of hydrogen atoms in water or ice.

The concentration n in (3.7) can be rewritten as $n = N_{eff}/V_0$, where N_{eff} is the number of charge carriers per volume V_0 of a single H_2O molecule. Then, one can write

$$N_{eff}(\omega) = \frac{2mV_0}{\pi q^2} \int_0^{\omega} \sigma(\omega')d\omega'. \tag{3.8}$$

The function N_{eff} is shown on the right-hand scale of Fig. 3.10 as a function of the cutoff frequency. The values of N_{eff} coincide for ice and water at two points marked with the arrows 1 and 2 and also equal each other at frequencies above arrow 2, where $S(\omega_0) = S_\infty = \text{const}$, and $N_{eff} \approx 1$ (one conducting proton per H_2O molecule). At frequencies above arrow 1, the spectra are dominated by intramolecular vibrations, as evident from a comparison of the H_2O and D_2O (see Sect. 2.7). On the contrary, the intermolecular charge motion dominates below arrow 1, or in the region where dielectric relaxation occurs. The transition point between intramolecular and intermolecular dynamics lies near the 5 THz vibration mode (see Sect. 2.6.1), where $N_{eff} \approx 0.02$ (see Fig. 3.9), or, in other words, corresponds to 1 proton charge per 50 H_2O molecules. Such a high concentration of charge carriers assumes a strong electrostatic interaction between them, because the average distance between charges of about 2 nm [25] is shorter than the corresponding Debye screening length.[6] Thus, the smallest "quantum" of the intermolecular conductivity of ice and water is a proton (or proton hole) charge shared between 50 water molecules. Excess protons, or in other words ions of H_3O^+ and OH^- obey Brownian motion in the potential of mutual screening, caused by electrostatic interaction.

3.5 Protonic Transport as a Fundamental Mechanism of the Dielectric Response of Ice and Water

Let us now consider the oscillatory-diffusion motion of a single excess proton (an excited third proton on the water molecule that briefly forms an H_3O^+ ion) of a mass m in a time-dependent potential (or in a polarizable atmosphere), taking into account its electrostatic interaction with other excess protons and proton holes (the missing proton on the water molecule that briefly forms an OH^- ion). Hereafter, we do not distinguish between excess protons and proton holes, unless otherwise stated, because the mechanism of their diffusion is nearly the same. Both H_3O^+ and OH^- obey Brownian-like diffusion as a whole between spontaneous proton-transfer events by the Grotthuss mechanism (see Sect. 1.3), and electrostatically interact with each other. Figure 3.11 shows a schematic snapshot of the corresponding water structure. Each charge has two polarization spheres. The first sphere (the dashed black circle) is formed by polar H_2O molecules. The second one (the dashed red circle) is made of charges of the opposite sign.[7] Each excess proton can move inside the ionic atmosphere formed by the surrounding negative charges (proton holes) and together with its center. The latter dynamics are expected to be responsible for static conductivity, while the former is responsible for the high-frequency polarization effects. Such an approach to the microscopic description of the dielectric response allows one to reduce the problem of polarization in liquid water and ice to a single-particle task.

---

[6]The Debye length is the distance over which the electric field of a separate charge, placed in a quasi-neutral medium containing free positively and negatively charges, becomes significantly screened.

[7]The optimal distribution of charge in an electrically neutral system of charges of two types is when positively charged ions are surrounded by negatively charged ones and vice versa.

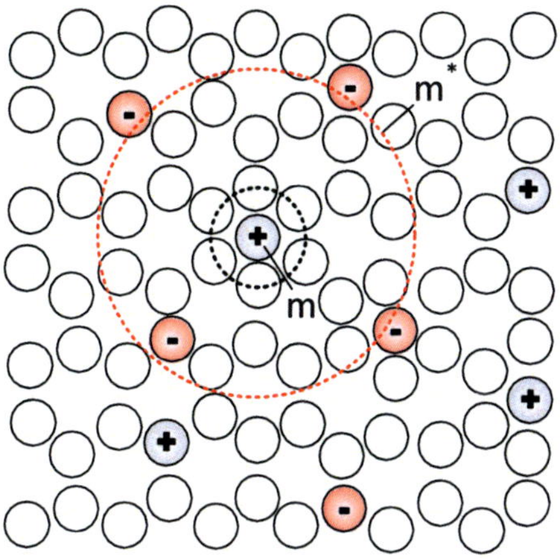

Fig. 3.11 A schematic representation of water's "structure." The charge in the center represents a short-lived H_3O^+ ion (violet), which is surrounded by short-lived OH^- ions (red). There are two polarization spheres: (1) the ionic atmosphere (the dashed red circle), and (2) the hydration shell of the neutral polar H_2O molecules (the dashed black circle). The dynamics of the central ion and its ionic atmosphere are described by (3.9) and (3.10), respectively

Using the general theory of the dynamics of interacting Brownian particles [41], we write a single-particle Langevin equation for a proton inside the polarization atmosphere (in-cage motion) [42]:

$$m\ddot{x} + m\gamma\dot{x} + \kappa_2 \int_0^t M(t - t')\dot{x}(t')dt' + K = f(t), \tag{3.9}$$

where x is the position of the particle, γ is a damping constant, $\kappa_2 = \mathrm{m}\omega_0^2$ is a coupling constant with ω_0 being the effective frequency of the oscillations, K is a restoring force due to the interaction with the center of the atmosphere, and $f(\mathrm{t})$ is a random force. The memory function, $M(\mathrm{t{-}t'}) = \exp[-(\mathrm{t{-}t'})/\tau_c]$, expresses the frequency-dependent damping with τ_c being the critical transition time from in-cage motion to long-range diffusion motion [41].

The polarization atmosphere of an excess proton, in turn, obeys Brownian motion according to

$$m^*\ddot{X} + m^*\Gamma\dot{X} - K = F(t), \tag{3.10}$$

where X is the center of the atmosphere, m^* is the effective mass (see Fig. 3.11), Γ is the damping constant, $F(\mathrm{t})$ is a stochastic force, $K = \kappa_1(x - X)$ is the restoring force (the same as in (3.9)), and $\kappa_1 = \mathrm{m}\Omega_0^2$ with Ω_0 being a measure of the restoring force.

According to the fluctuation–dissipation theorem [44], the frequency-dependent conductivity, $\sigma(\omega)$, is defined by the velocity–velocity autocorrelation function, $< \dot{x}(t)\dot{x}(0)>$, by the formula:

$$\sigma^*(\omega) = \frac{n_i q}{k_B T} \int < \dot{x}(t)\dot{x}(0) > e^{i\omega t} dt. \quad (3.11)$$

Using standard formalism [42], (3.9) and (3.10), give the following solution [43]:

$$\sigma^*(\omega) = \frac{i n_i q^2 B}{(\kappa_1/\omega)^2 - AB}, \quad (3.12)$$

with

$$A = \frac{\kappa_1}{\omega} - m[\omega + i(\gamma + \omega_0^2 M(\omega)], \quad (3.13)$$

and

$$B = \frac{\kappa_1}{\omega} - m^*\omega + i \cdot m^*\Gamma, \quad (3.14)$$

where $M(\omega) = 1/(\tau_c - 1 - i\omega)$ is a Laplace transform of the memory function $M(t)$.

Separating the real and imaginary parts in (3.12), we get

$$\sigma'(\omega) = \frac{n_i q^2 \omega}{m\tau_c} \frac{D(E - \omega^2\Gamma\tau_c) - C\omega(\Gamma + E\tau_c)}{C^2 + D^2}, \quad (3.15)$$

and

$$\varepsilon'(\omega) = \varepsilon_\infty + \frac{n_i q^2}{m\varepsilon_0\tau_c} \frac{C(E - \omega^2\Gamma\tau_c) + D\omega(\Gamma + E\tau_c)}{C^2 + D^2}, \quad (3.16)$$

where

$$C = FE - G\omega\Gamma - \frac{m}{m^*}\frac{\Omega_0^4}{\tau_c}, \quad (3.17)$$

$$D = GE + F\omega\Gamma - \omega\frac{m}{m^*}\Omega_0^4, \quad (3.18)$$

$$E = \frac{m}{m^*}\Omega_0^2 - \omega^2, \quad (3.19)$$

Table 3.5 The parameters used to fit (3.12) to the experimental spectra of water and ice at 273 K shown in Figs. 3.8 and 3.9. Additionally, $m_p = 1.67 \cdot 10^{-27}$ kg, $q = 1.6 \cdot 10^{-19}$ C, $\epsilon_\infty = 2.5$ and 2.2 for ice and water, respectively

	$n_i \times 10^{26}$ (m^{-3})	m^*/m_p	γ (THz)	Γ (THz)	$\omega_0/2\pi$ (THz)	$\Omega_0/2\pi$ (THz)	τ_c (ms)
Ice	7.7	18	2.9	$2.4 \cdot 10^6$	0.6	5.4	5.1
Water	8.2	18	4.3	2.2	0.7	5.0	0.6

$$F = \frac{\Omega_0^2 - \omega^2(1 + \gamma\tau_c)}{\tau_c}, \tag{3.20}$$

$$G = \omega\left(\Omega_0^2 - \omega^2 + \omega_0^2 + \gamma/\tau_c\right). \tag{3.21}$$

Equations 3.15 and 3.16 with coefficients defined by (3.17)–(3.21) are analytical expressions for the frequency-dependent conductivity (compare with 2.25 and 2.26) that include oscillatory and diffusion parts, and two types of interaction: the short-range friction between a proton and its nearest neighbors and the long-range mutual Coulomb interaction between excess protons. Inasmuch as all excess protons and proton holes are electrostatically connected, the damping Γ is related to the collective rearrangements or ambipolar diffusion[8] of these charges [45].

Best fits of (3.15) and (3.16) to the dielectric spectra of ice and water are shown in Fig. 3.8 (solid black lines); the fit parameters are given in Table 3.5 (compare with the much larger number of parameters in Table 2.1 for other fits). One can see the excellent agreement of the model and the experimental data up to 10 THz for both ice and water. The model, given by (3.12), simultaneously describes the DC conductivity σ_{dc}, the high-frequency conductivity σ_∞, and the infrared intermolecular vibrations. It also fulfills the sum rule, given by 2.61.

For DC conductivity (3.15) gives

$$\sigma_{dc} = \frac{n_i q^2}{m\gamma + \kappa_2\tau_c + m^*\Gamma} \approx \frac{n_i q^2}{\kappa_2\tau_c}. \tag{3.22}$$

This equation shows that the static conductivity of both ice and water are products of the ambipolar diffusion of interacting charge carriers. One can see the inverse proportionality of σ_{dc} to the transition time τ_c and the coupling constant $\kappa_2 = m_p\omega_0^2 = 0.032$ N/m.

Note that τ_c is the time required for the excess proton to change its neighboring ions completely or move out from the polarization atmosphere, thus showing DC

[8] Ambipolar diffusion is the diffusion of positive and negative species with opposite electrical charge due to their interaction via an electric field.

conductivity. The corresponding barrier as evident from the analysis of the temperature dependencies of conductivity and dielectric relaxation in water [25] is a fraction of an eV. With this barrier ΔE, the time τ_c is equal to

$$\tau_c = \frac{1}{\nu_0} \exp\left(\frac{\Delta E}{k_B T}\right). \tag{3.23}$$

Equation 3.23, even at a high frequency of charge oscillations, $\nu_0 \approx 1$ THz, takes the value of about one millisecond (see Table 3.5). This is a result of cooperativity in the protonic "superlattice," formed by an ensemble of excess protons.

For the dielectric constant (3.16) gives

$$\varepsilon(0) = \frac{n_i q^2 (m^* \Gamma + \kappa_1 \tau_c)}{\varepsilon_0 \kappa_1 (m\gamma + \kappa_2 \tau_c + m^* \Gamma)} \approx \frac{n_i q^2}{\varepsilon_0 \kappa_2}. \tag{3.24}$$

One can see that $\epsilon(0)$ depends on the concentration n_i and the coupling constant κ_2. Introducing the plasma frequency, $\omega_p^2 = n_i q^2 / \epsilon_0 m$, we have $\epsilon(0) = \omega_p^2 / \omega_0^2$ from (3.24). Thus, within this model $\epsilon(0)$ is formed similar to plasma by the relative displacements of the excess protons and the holes. The same microscopic mechanism, as dictated by the causality principle, applies for dielectric relaxation phenomena in both ice and water.

Table 3.5 shows that the damping constants, Γ, for ice and water differ significantly, showing the presence of the long-range order in ice formed by ions and its impossibility in liquid water presumably due to the fast proton exchange between ions and H_2O molecules. The lifetime of water molecules, obviously, depends on the charge hoping rate. The high damping constant Γ for ice is a result of the slow rate of proton exchange (in comparison with water) and the formation of long-range order by analogy with ionic crystals. Other parameters, namely, the concentration of conducting species n_i, the damping constant γ, the effective mass m^*, and frequencies Ω_0 and ω_0, remain stable.

The fact that two sets of microscopic parameters in Table 3.5 coincide indicates the similar structures of ice and water, with the only difference being the speed of proton dynamics. The effective mass, $m^* = 18 \cdot m_p$, in the model equals the mass of H_3O^+ and OH^- ions, because the excess proton is always connected to a host molecule. The frequency $\Omega_0 = 5$ THz (180 cm^{-1}) equals the frequency of mode 2 (Table 3.4).

Because both (3.6) and (3.12), fit the experimental spectra, one can conclude that mode 2 corresponds to the vibration of the ions and mode 1 is the diffusion of ions in the mutual electrostatic interaction potential. The corresponding 95% of contribution of mode 1 to the static dielectric constant $\epsilon(0)$ is due to the proton-hole relative polarization.

The concentration n_i of excess protons (see Table 2) corresponds to 4% of the total concentration of water molecules, $n_w = 3 \cdot 10^{28}$ m^{-3}. At first glance, this concentration contradicts to the concept of pH. However, the pH is determined by the static conductivity and with the assumption that ions in water do not interact. In the considered model, n_i is obtained assuming electrostatic interaction among ions. Thus, the pH is a measure of proton activity (but not of the concentration of excess protons) at the static limit only. The majority of these excess protons sit in the potential and conduct only at high frequencies.

The high-frequency ion diffusion coefficients,

$$D_\infty = \frac{\sigma_{D1} k_B T}{n_i q^2}, \tag{3.25}$$

are equal $1.1 \cdot 10^{-14}$ for ice and $7 \cdot 10^{-9}$ m^2/s for water. They exceed the self-diffusion coefficient of H_2O molecules by 10 and 350% for ice and water, respectively. This means that the excess protons (or holes) in liquid water do not remain bound to water molecules and can spontaneously change hosts by proton exchange. As a consequence, the ensemble of excess protons forms a kind of gas, if we switch to the frame of reference of a neutral water molecule and a crystal-like structure in a laboratory frame of reference. Note that the charge carrier here is not equivalent to the hydrogen proton, as a water molecule has two hydrogen protons and either of them can participate in the charge transfer.

It should be noted that we do not explicitly discuss the hydrogen bonds between the H_2O molecules in this model. However, the manifold of the excess-proton migration and the corresponding dissolution of the excess protons among neutral water molecules on large time and space scales can be considered as an analogue of hydrogen bonding (see Sect. 4.1).

The spectra of water and ice in terms of the real, ϵ', and imaginary, ϵ'', parts of the complex dielectric function $\epsilon^*(\nu) = \epsilon'(\nu)+i\epsilon''(\nu)$ have fundamental similarities discussed in Sect. 3.4. The general difference between the two spectra is a frequency shift shown by the arrow in Fig. 3.9. The static dielectric constant ($\epsilon(0)$) is shown in Fig. 3.12 as a function of temperature. The experimental curve $\epsilon(0, T)$ looks like a single line with a small gap near the melting point substantiated, presumably, by the 8% density change between ice and water.[9] Because $\epsilon(0)$ is proportional to the integral of $\epsilon''(\nu)$, considering the similar mechanism for ice and water dielectric functions, we expect similar mechanisms for their dielectric constants.

From (3.23) and (3.24), one can obtain for the static dielectric constant:

$$\varepsilon(0) = \frac{\sigma_{dc} \tau_c}{\varepsilon_0}. \tag{3.26}$$

This relation links $\epsilon(0)$ with DC conductivity σ_{dc} and is proportional to the transition time τ_c.[10] The DC conductivity and τ_c are both functions of temperature:

[9] Note that the difference between the concentrations of excess protons in ice and water in Table 3.5 is also about 8%.

[10] Note that this formula is different from the classic Debye formula $\epsilon(0) = \sigma_{D1} \tau_{D1}/\epsilon_0$, which connects high-frequency conductivity σ_{D1} and dielectric relaxation time τ_{D1} with the static dielectric constant.

Fig. 3.12 The temperature dependence of the static dielectric permittivity, $\epsilon(0)$, of ice (blue) and water (black). The orange line is a fit according to the (3.28). Data from [3]

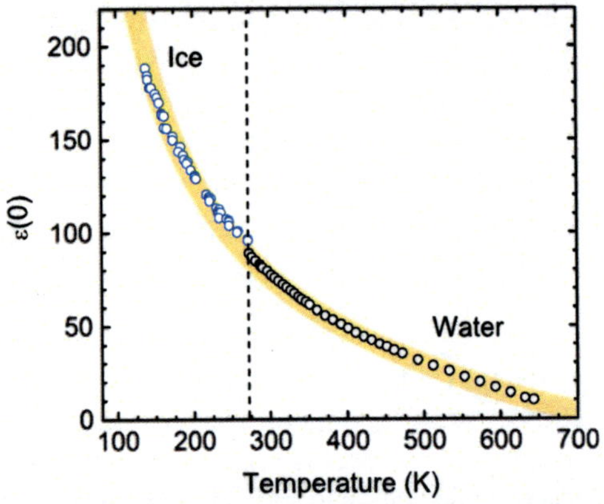

$$\sigma_{dc} = \frac{\sigma_0}{T} \exp\left[\frac{-E_a^{\sigma}}{k_B(T - T_0)}\right], \tag{3.27}$$

$$\tau_c = \frac{1}{\omega_0} \exp\left[\frac{E_a^{\tau}}{k_B(T - T_0)}\right], \tag{3.28}$$

where E_a are two different activation energies, ω_0 is the oscillation frequency defined above, and T_0 is the critical temperature. Substitution of (3.27) and (3.28) for (3.26) gives the static dielectric constant:

$$\varepsilon(0, T) = \frac{A}{T} \exp\left[\frac{E_a}{k_B(T - T_0)}\right], \tag{3.29}$$

where $A = \sigma_{dc}/(\epsilon_0\omega_0)$ and $E_a = E_a^{\tau} - E_a^{\sigma}$. Equation 3.29 is a single relation for $\epsilon(0)$ for both ice and water. It does not contain the H_2O dipole moment μ_0 and perfectly fits the experimental $\epsilon(0)$ of ice and water with the following parameters: $A = 5 \cdot 10^4$ K, $E_a = 0.04$ eV, and $T_0 = 840$ K, as shown in Fig. 3.12. Interestingly, at $T = T_0$, the dielectric constant $\epsilon(0)$ reaches $\epsilon_\infty \approx 3$. At $T \ll T_0$, (3.29) gives a Curie-like law, $\epsilon(0) \sim C/T$, with $C = 2 \cdot 10^4$ K.

Thus, the polarization due to local fluctuations of ions by the spontaneous exchange of excess protons with neutral molecules is an appealing alternative to the rotating dipoles of H_2O. Note that the suggested model is applicable for a wide frequency range without any additional species except intrinsic ions. This is a characteristic feature of the model.

The model produces a set of parameters that shows the identity of the local dynamics in ice and water (see Table 3.5). This shows that the proton exchange rate is the only parameter that is different for ice and water. It governs the formation of a long-range network of electrostatically connected ions. Fast proton exchange in water makes the formation of the long-order structures, reflected in the model by the damping constant, impossible.

The proposed model is a single formula (3.12) instead of the two relaxations and four oscillators, commonly used for the same spectral region [46, 47]. The number of parameters is reduced from $3\times6 = 18$ to 6. Each parameter has a clear physical meaning. The model explains not only the polarization mechanism, but also DC conductivity. This is a distinctive feature of the model compared to existing models [20, 48, 49]. In addition, it treats ice and water on the same footing. These three advantages extend the range of the validity of the model.

We analyzed ice and water dielectric responses in an external electric field on the same footing. We introduced a phenomenological model based on an analogy between ionic conductors, ice and water. The model shows good agreement with experimental data over fourteen orders in frequency magnitude, predicts the general trend of the temperature dependence of the static dielectric constant, and is useful for experimental data approximations. According to this model, ice and water conduct electricity by the interaction of excess protons and proton holes. No other defects, except short-living H_3O^+ and OH^- ions, are required for the dielectric-spectra interpretation from DC up to terahertz. The model sheds new light on the electrical properties of water, which were not fully accounted for by previous models [20, 50]. It also provides a simple means for analytical ice and water dielectric data approximation under a wide range of thermodynamic conditions within an extended frequency range, up to 10^{13} Hz.

References

1. K. Steffen, S.V. Nghiem, R. Huff, G. Neumann, The melt anomaly of 2002 on the Greenland Ice Sheet from active and passive microwave satellite observations. Geophys. Res. Lett. **31**, L20402–5 (2004)
2. V.F. Petrenko, R.W. Whitworth, *Physics of Ice* (University Press, Oxford, 1999)
3. V.G. Artemov, A unified mechanism for ice and water electrical conductivity from direct current to terahertz. Phys. Chem. Chem. Phys. **21**, 8067–8072 (2019)
4. P.V. Hobbs, *Ice Physics* (Clarendon Press, Oxford, 1974)
5. A. von Hippel, The dielectric relaxation spectra of water, ice, and aqueous solutions, and their interpret at ion 1. Critical survey of the status-quo for water. IEEE Tnans. Electr. Insul. **23**, 801–816 (1988)
6. I. Takei, Physics and Chemistry of Ice, in *Proceedings of the 11th International Conference on the Physics and Chemistry of Ice* ed. by Kuhs WF, vol 430 (2007)
7. S.G. Warren, R.E. Brandt, Optical constants of ice from the ultraviolet to the microwave: a revised compilation. J. Geoph. Res. **113**, 10–D14220 (2008)
8. T. Matsuoka, S. Fujita, S. Mae, Effect of temperature on dielectric properties of ice in the range 5–39 GHz. J. Appl. Phys. **80**, 5884–5890 (1996)
9. D.C. Elton, M. Fernandez-Serra, The hydrogen-bond network of water supports propagating optical phonon-like modes. Natl. Commun. **7**, 10193–8 (2016)
10. D.D. Klug, E. Whalley, Origin of the high-frequency transkational bands of ice I. J. Claciol. **21**, 55–63 (1978)
11. G.E. Walrafen, Raman spectrum of water: transverse and longitudinal acoustic modes below $\approx 300\ cm^{-1}$ and optic modes above $\approx 300\ cm^{-1}$. J. Phys. Chem. **94**, 2237–2239 (1990)
12. K. Abe, T. Shigenari, Raman spectra of proton ordered phase XI of ICE I. Translational vibrations below $350\ cm^{-1}$. J. Chem. Phys. **134**, 104506–11 (2011)

13. J.E. Bertie, H.J. Labbe, E. Whalley, Absorptivity of Ice I in the range 4000–30 cm^{-1}. J. Chem. Phys. **50**, 4501–4520 (1969)
14. E. Whalley, D.D. Klug, Comment on TO-LO splitting in disordered phases of ice. J. Chem. Phys. **71**, 1513–1513 (1979)
15. T.S. Light, S. Licht, A.C. Bevilacqua, K.R. Morash, The fundamental conductivity and resistivity of water. Electrochem. Solid-State Lett. **8**, E16–E19 (2005)
16. R. Mills, Self-diffusion in normal and heavy water in the range 1–45°. J. Phys. Chem. **77**, 685–688 (1973)
17. A.J. Easteal, W.E. Price, L.A. Woolf, Diaphragm cell for high-temperature diffusion measurements. Tracer Diffusion coefficients for water to 363 K. J. Chem. Soc. Faraday Trans. **1**, 1091–1097 (1989)
18. M. Holz, S.R. Heil, A. Sacco, Temperature-dependent self-diffusion coefficients of water and six selected molecular liquids for calibration in accurate ^{1}H NMR PFG measurements. Phys. Chem. Chem. Phys. **2**, 4740–4742 (2000)
19. N. Bjerrum, Structure and properties of ice. Del Kongelige Danske Videnskabernes Selskab. Matematisk-fysiske Meddelelser. **27**, 1–56 (1951)
20. I. Popov, P. Ben Ishai, A. Khamzin, Y. Feldman, The mechanism of the dielectric relaxation in water. Phys. Chem. Chem. Phys. **18**, 13941–13953 (2016)
21. H. Gränicher, Gitterfehlordnung und physikalische Eigenschaften hexagonaler und kubischer Eiskristalle. Z. Kristallogr. **110**, 432–471 (1958)
22. C. Jaccard, Étude théorique et expérimentale des propriétés électriques de la glace. Helv. Phys. Acta. **32**, 89–128 (1959)
23. R. Podeszwa, V. Buch, Structure and dynamics of orientational defects in ice. Phys. Rev. Lett. **83**, 4570–7573 (1999)
24. L. Onsager, L.K. Runnels, Mechanism of self-diffusion in ice. PNAS **50**, 208–210 (1963)
25. V.G. Artemov, A.A. Volkov, A.V. Pronin, Electrical properties of water: a new insight. Biophysics **59**, 520–523 (2014)
26. V.G. Artemov, A. Ryzhov, E. Carlsen, P.O. Kapralov, H. Ouerdane, Nonrotational mechanism of polarization in alcohols. J. Phys. Chem. B. **124**, 11022–11029 (2020)
27. P. Debye, H. Falkenhagen, Dispersion der Leitfähigkeit starker Elektrolyte. Phys. Z. **34**, 562–565 (1928)
28. J. O. Bockris, A. Reddy, *Modern Electrochemistry* 2nd edn. (Kluwer, 2002)
29. K.V. Rao, Electrical conductivity and dielectric properties of NaCl, KCl, and KBr single crystals near their melting temperatures. Physica Status Solidi (a) **30**, 391–395 (1975)
30. M.-M. Huang, Y. Jiang, P. Sasisanker, G.W. Driver, H. Weingärtner, Static relative dielectric permittivities of ionic liquids at 25 °C. J. Chem. Eng. Data. **56**, 1494–1499 (2011)
31. S. Bhattacharya, A. Ghosh, Relaxation of silver ions in superionic borate glasses. Chem. Phys. Lett. **424**, 295–299 (2006)
32. C. Filipič, A. Moguš-Milanković, L. Pavić, K. Srilatha, N. Veeraiah, A. Levstik, Polaronic behavior of MnO doped LiI-AgI-B_2O_3 glass. J. Appl. Phys. **112**, 073705–3 (2012)
33. P. Atkins, J. de Paula, *Physical Chemistry*, 9th edn. (W. H. Freeman and Company, New York, 2010)
34. V.G. Artemov, I.E. Kuritsyna, S.P. Lebedev, G.A. Komandin, P.O. Kapralov, I.E. Spektor, V.V. Kharton, S.I. Bredikhin, A.A. Volkov, Analysis of electric properties of ZrO_2-Y_2O_3 single crystals using teraherz IR and impedance spectroscopy techniques. Russ. J. Electroch. **50**, 690–693 (2014)
35. M. Fox, *Optical Properties of Solids*, 2nd edn. (University Press, Oxford, 2010)
36. H. Fröhlich, *Theory of Dielectrics*, 2nd edn. (University Press, Oxford, 1958)
37. K.F. Young, H.P.R. Frederikse, Compilation of the static dielectric constant of inorganic solids. J. Phys. Chem. Ref. Data. **2**, 313–409 (1973)
38. P. Lunkenheimer, V. Bobnar, A.V. Pronin, A.I. Ritus, A.A. Volkov, A. Loidl, Origin of apparent colossal dielectric constants. Phys. Rev. B. **052105–4** (2002)
39. I.A. Ryzhkin, Ice in Proton conductors, in *Solids, Membranes and Gels Materials and Devices*, ed. by P. Colomban (Cambridge University Press, Cambridge, 1992), pp. 158–164

40. V.G. Artemov, Dielectric spectrum of water as a proton dynamics response. Bull. Lebedev Phys. Inst. **42**, 187–191 (2015)
41. W. Dieterich, I. Peschel, Memory function approach to the dynamics of interacting Brownian particles. Physica A: Stat. Mech. Appl. **95**, 208–224 (1979)
42. R. Zwanzig, *Nonequilibrium Statistical Mechanics* (Oxford University Press, Oxford, 2001)
43. P. Fulde, L. Pietronero, W.R. Schneider, S. Strässler, Problem of brownian motion in a periodic potential. Phys. Rev. Lett. **35**, 1776–1779 (1975)
44. R. Kubo, The fluctuation-dissipation theorem. Rep. Prog. Phys. **29**, 255–284 (1966)
45. L.D. Landau, E.M. Lifshitz, *Physical Kinetics* (Pergamon Press, New York, 1981)
46. H. Yada, M. Nagai, K. Tanaka, Origin of the fast relaxation component of water and heavy water revealed by terahertz time-domain attenuated total reflection spectroscopy. Chem. Phys. Lett. **464**, 166–170 (2008)
47. R. Buchner, J. Barthel, J. Stauber, The dielectric relaxation of water between 0 °C and 35 °C. Chem. Phys. Lett. **306**, 57–63 (1999)
48. A.A. Volkov, V.G. Artemov, A.V. Pronin, A radically new suggestion about the electrodynamics of water: can the pH index and the Debye relaxation be of a common origin? EPL **106**, 46004–6 (2014)
49. C. Jaccard, Mechanism of the electrical conductivity in ice. Ann. NY Acad. Sci. **125**, 390–400 (1965)
50. M.J. Gillan, D. Alfé, A. Michaelides, Perspective: How good is DFT for water? J. Chem. Phys. **144** (2016)

Chapter 4
The Dielectric Properties and Dynamic Structure of Water and Ice

Abstract The electrodynamic properties of water are inextricably linked with its structure and molecular dynamics. This chapter discusses a phenomenological approach that provides a microscopic description of the various electrodynamic parameters of water and ice on the same basis. The parameters related to the microscopic dynamics of water (such as the molecular lifetime, the rate of proton exchange, autoprotolysis, and the mechanism of dielectric polarization) are discussed in fine details and compared with those derived from independent experimental data. The microscopic description of the phenomenon of dielectric relaxation and an explanation for the anomalous behavior of dielectric constant are given. In addition, questions on why microwaves are absorbed by water and pass through other dielectrics, and why pH is so dependent on temperature are addressed.

4.1 Problems of Describing the Dynamics of Water on the Basis of Hydrogen Bonds

As mentioned in Chap. 1, hydrogen bonding is a phenomenological concept that provides a good platform for the qualitative description of a variety of water's anomalies.[1] However, although this concept allows one to explain key properties of water, the vague definition [1], and multiple parameters that are used for the parametrization of the bonds, do not assume a simple microscopic picture for the atomic-molecular dynamics, hindering a quantitative description of the majority of the (electro)dynamic parameters of water and ice. A special problem is the abundance of ambiguous descriptions of the parameters of the hydrogen bonds, such as shape, lifetime, length, directionality, and energy. In general, the concept of "bond" is probably not the best way to describe the molecular dynamics, as it does not account for the exchange of atoms between molecular species or the fact that electrostatic interactions are a function of distance, which is controlled by the thermal fluctuations.

[1] See a comprehensive description of the anomalies of water on the website of Prof. Martin Chaplin: http://www1.lsbu.ac.uk/water.

V. Artemov, *The Electrodynamics of Water and Ice*, Springer Series in Chemical Physics 124, https://doi.org/10.1007/978-3-030-72424-5_4

The complexity of the concept of hydrogen bonding is obvious from the fact that the notion of a hydrogen bond includes many types of interactions, including electrostatic attraction, polarization, dispersion, proton exchange, and covalent bonds,[2] and that it has no strict criteria that allow one to "recognize" the hydrogen bond among other molecular interactions [1]. The criteria are still too blurry, and the relative contribution of the individual components of hydrogen bonds that keep the species of water together remains a matter of debate [2, 3].

The specificity of bonds between water molecules can be demonstrated in a simple thought experiment. The amount of energy necessary for the evaporation of water at normal pressure is $Q = 2260$ kJ/kg [4]. Assuming that water consists of H_2O species,[3] one gets their molar concentration $n = m/M = 1000\,\text{g}/18\,\text{g} \cdot \text{mol}^{-1} = 55.5\,\text{mol/l}$, where m is the mass, and M is the molecular mass. Dividing Q by n, one gets the energy per molecule, which equals $q_0 = 40.7$ kJ/mol. Assuming that on average there are 3.6 "bonds" per molecule [5], one gets the energy of a single bond between two water molecules: $q_{HB} \approx 11.3$ kJ/mol. This bond is too weak for covalent or ionic bonds, which lie between 60 and 600 kJ/mol, and too strong for Van-der-Waals interactions, which lie between 0.1 and 4.0 kJ/mol [4]. Thus, the bonds between water molecules were assigned to a special type of intermolecular interactions, or hydrogen bonds [6]. The value q_{HB} is close to the value 13.6 kJ/mol of the binding energy of the water dimer $(H_2O)_2$ [7], and close to the spread of about 11 kJ/mol (0.11 eV) in the interaction energy among different isomers of the water dimer [3]. The latter energy is obviously a minimal energy barrier of molecular diffusion.[4] In such a way, the concept of hydrogen bonding found a logical framework and has spread widely across different fields of science [6]. However, it has significant limitations, which make it inapplicable at short (picosecond) timescales.

Note that the calculations provided above are based on the assumption that water is an ensemble of sharply defined H_2O species. The diffusion-averaged structure of water possess a molecular nature, however, at short (picosecond) timescales, simulations of molecular dynamics, and ultrafast spectroscopic techniques (see Chap. 1), show that molecules lose their individuality and can form the short-lived species which are missing in the diffusion-averaged structure of water by Bernal and Fowler (see Chap. 1). Each short-lived charged species, such as hydronium, Eigen, or Zundel cations, and their counterions (anions), are the centers of the Coulomb field, which electrostatically affects the surrounding molecules. These kinds of intermolecular interactions are missing in the conventional concept of hydrogen bonding. For instance, Eigen showed [8] that the heat of the hydration of the hydronium (H_3O^+) or hydroxyle (OH^-) ions of water is relatively large and is approximately 1200 kJ/mol

[2]Covalent bonding includes σ-bonding, π-bonding, metal-to-metal bonding, agostic interactions, bent bonds, three-center two-electron bonds, and three-center four-electron bonds.

[3]See the notes on the DC conductivity of water in Sect. 1.3, which show that this assumption is not totally correct, because the small concentration of intrinsic ions (H_3O^+ and OH^-) of water was obtained neglecting their mutual interaction. Accounting for the interaction leads to a much higher concentration of short-lived ionic species (see Sect. 3.5), which cannot be neglected when intermolecular interaction is taken into account.

[4]The experimental activation energy of self-diffusion in water is 0.18 eV (see Sect. 1.4).

(300 kcal/mol). This energy was not accounted for in the above-calculated strength of the hydrogen bond. Note that the electrostatic field of ions polarizes the surrounding species more efficiently than the dipole–dipole interaction and assumes long-range intermolecular cooperativity, because the electrostatic potential of an ion is proportional to the inverse distance $1/r$, but the Van-der-Waals potential is proportional to $1/r^6$. While the latter can be neglected at the intermolecular level, the former provides an additional cohesion of water species on the long-range level, which must be accounted for.

Eisenberg and Kauzmann [9] analyzed the problems of hydrogen bonding and concluded that the fraction of broken bonds in the liquid may not be the best parameter for describing water. For example, the IR spectrum, discussed in Sect. 2.5, is not compatible with the concept of a discrete number of sharply defined bounded species. If such species exist, one would expect a contribution from non-hydrogen-bonded OH groups of H_2O molecules near the corresponding OH-vibration mode of water vapor (see Fig. 2.13). However, this was not observed. Moreover, the large dielectric constant, $\epsilon(0)$, of water and ice implies strong molecular correlations and a high polarization ability. These two mechanisms cannot be simultaneously satisfied in terms of a hydrogen-bonded network of identical long-lived molecular species.

Although some aspects of the complex phenomenon of hydrogen bonding are still suitable for the description of proton transfer, in general, it is outdated and does not fully reflect modern spectroscopic data. The dynamics of hydrogen bonds and Bernal–Fowler water (see Chap. 1) is probably not the best platform for the analysis of the electrodynamics of water and ice. The concept of molecular water was developed on the basis of diffusion-averaged dynamics and does not include short-term (picosecond) high-frequency dynamics, i.e., high-frequency dynamics assume additional to H_2O short-lived species [10]. In Sects. 4.2 and 4.3 we discuss a phenomenological model (the ionic model) of water, in which a variety of specific bonds of well-defined molecular species are substituted with the simple electrostatic interaction between the variety of short-lived ions and molecules. Although the ionic model is yet to be tested and recognized, it looks promising for the explanation of the anomalous properties of water beyond the electrodynamic data.

4.2 A Phenomenological Model for the Broadband Dielectric Response

4.2.1 Microscopic Features of Self-diffusion

Frenkel showed [11] that molecules of a liquid undergo Brownian diffusion, being at the same time in an oscillatory state similar to that in solids. In this way, he reduced the theoretical gap between solids and liquids, allowing them to be considered on the same footing. Figure 4.1 shows a snapshot of the relative arrangement of water molecules estimated by neutron and X-ray diffraction techniques (see Sect. 1.2.3 for

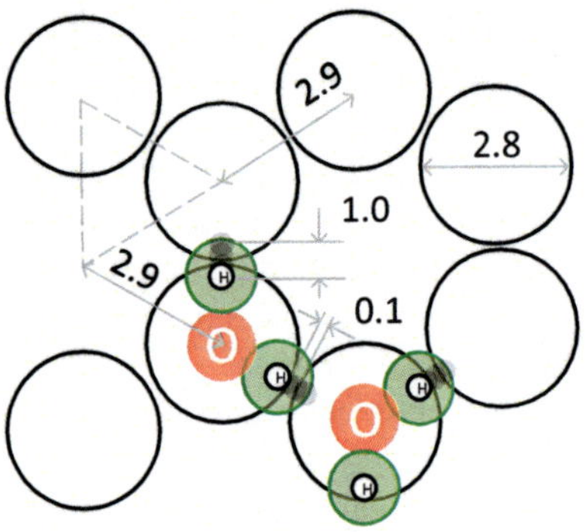

Fig. 4.1 The local molecular environment in water. The black circles outline the contours of the water molecules, the green circle represents the area of the delocalization of protons, which partially overlaps with the nearest molecule. The arrows and numbers show the corresponding dimensions in Å

details). This microscopic structure is characterized by the oxygen atoms forming hexagonal rings with a nearly tetrahedral coordination of molecules. The average distance between nearest oxygen atoms is about 2.9 Å. The water molecule packing density is 0.36 [12], which is half the density of 0.74 for the close-packing of spherical particles. Thus, the volume of cavities between the molecules is approximately the same as the occupied volume, which provides enough space for molecular diffusion.[5] The diameter of a water molecule is 2.8 Å, the average distance between molecules (the amplitude for Frenkel's oscillations) is about $d_0 = 0.1$ Å, and the distance to the center of nearest cavity is $l_0 = 2.9$ Å. Thus, there are several possible degrees of freedom for molecular diffusion: molecular oscillations, orientations, interstitial translations, and independent diffusion of oxygen and hydrogen atoms by proton exchange.

There is no direct experimental evidence that hydrogen atoms (protons) are pinned to the oxygen atoms for a long period of time. The lifetime of water molecules was initially established by conductivity measurements; the experiment's pros and cons are discussed in Sect. 1.3. For example, eigen thought that the average period of association of proton with a water molecules is less than 1 ps [8], while Bernal and Fowler talk about 11 h [13]. The de Broglie wavelengths[6] of protons at room temperature is 1.8 Å [14], which significantly exceeds both the intermolecular gap d_0, and the width of the barrier of the proton transport reaction $\Delta x \approx 1$ Å [15]. Thus, the proton is essentially delocalized, and there is a high probability of proton exchange between molecules, which is the primary mechanism of the electrical conductivity

[5] After dissolving salt in water there is a reduction in volume due to the filling of the non-occupied volume.

[6] A de Broglie wave is a probability wave, which determines the probability density of object's detection at a given point in the configuration space. The notion is a central part of the theory of quantum mechanics, which represents the wave–particle duality.

of water.[7] In other words, the finite lifetime of H_2O molecules depends on the rate of the generation and recombination of excess protons (H_3O^+) and proton holes (OH^-).

The de Broglie wavelength is inversely proportional to particle momentum [14], thus, tunneling becomes noticeable in small masses and at low temperatures. However, if the process is diffusion controlled (as in Frenkel's concept of diffusion), the conditions change, and tunneling during hydrogenation/deuteration is accompanied by a semi-classical kinetic isotope effect with a negligible difference between the transfer probability of hydrogen and deuterium. Hama et al. [14] suggest that for this reason, the tunneling, which may account for the unexplained fast dynamics in water and aqueous systems, has been unrecognized in a variety of experimental studies. This, in particular, may apply to the relaxation part of the water spectrum below 10 THz, which demonstrates no significant isotopic shift after deuteration (see Sect. 2.7).

Bakker et al. [17] showed by pump–probe spectroscopy that the vibrational potential of the O–H stretch vibrations of water has extreme anharmonicity that arises from the O–H$\cdots$O interaction, and they observed that the energy required for this delocalization of the proton between the oxygen atoms of two neighboring water molecules is unexpectedly low. They found that the dissociation energy of the O–H bond of the water molecule in liquid water is at least 20% lower than that in the water vapor, which explicitly assumes the quantum-mechanical nature of the proton transfer. It was later confirmed [18] that Frenkel's thermally activated motion of water molecules leads to pronounced structural fluctuations, underlying processes ranging from sub-100 fs librational motions to picosecond intermolecular transformations. This timescale is in line with the results by Eigen [8], who also studied the dynamics at the same time interval.

Using ultrafast IR spectroscopy, Carpenter et al. [19] showed that the transport of an excess proton is followed by the reconstruction of its hydration shell,[8] which goes through the orientational relaxation of molecular dipoles within $\tau_{rot} \approx 2.5$ ps. The authors attributed the slow decay of the hydrated proton anisotropy to the proton intermolecular transfer kinetics that result in the loss of the memory of the prior configuration within the τ_{rot}. In other words, the hydration shell of the H_3O^+ ion relaxes to the equilibrium conditions after the perturbation caused by the intermolecular proton transfer in a picosecond time interval. The time decay τ_{rot} corresponds

[7]Interestingly, proton-transfer reactions have been found dramatically less common in simulations based on an empirical force field [16]. The problem is that extreme fluctuations of the O–H stretching vibrations are stabilized by the rearrangement of the electron density. This rearrangement is usually unaccounted for by the empirical potentials, which are defined by the X-ray diffraction of molecules of water vapor, where proton transfer is extremely rare. As a result, autoprotolysis is usually excluded from standard simulations of molecular dynamics (see Sect. 1.6.). However, this does not mean that the process is impossible in reality.

[8]The hydration shell is the layer of water molecules around the ionic species with perturbed structure and dynamics. The hydration shell provides an additional polarization degree of freedom in water and aqueous solutions.

to the second relaxation of water (see Sect. 2.6.3), thus, indirectly confirming the connection of the excess wing of the Debye relaxation to the orientational dynamics of H_2O dipoles in the hydration shell of short-lived ionic species.

While the short-range dynamics are diffusion controlled, the large-amplitude intermolecular proton dynamics are better described by the Grotthuss mechanism (see Sect. 1.3.1), in which the excess proton moves from one molecule to another by minimal displacements of individual H_2O species,[9] while the charge each time hops a larger distance, equal to the molecular diameter. Hassanali et al. [20] showed that rather than undergoing an exclusive step-wise hopping, the excess proton goes through periods, when it can cover 4–8 Å in distance over several H_2O molecules on sub-picosecond timescales, followed by a relatively long resting time τ_r, when the proton does not propagate through the water network, obeying Frenkel's oscillatory motion together with the host molecule. The authors showed that the excess proton resting time equals $\tau_r \geq 10$ ps, and that the correlated back-and-forth vibrations of protons between the neighboring water molecules exist on femtosecond timescales. The latter defines the lifetime of the Zundel complex.

Thus, in addition to Frenkel's idea of the simultaneous oscillatory-diffusional motion of molecules in a liquid, for water, we add the exchange of atoms by molecules. In other words, water differs from other liquids by the ability of molecules to exchange protons, which means that the species of water are short-lived and continuously being reconstructed.[10] The two types of atoms (oxygen and hydrogen) provide the formation of the following species in water: H_2O, H_3O^+, OH^-, and their derivatives: $H_3O^+{\cdot}3H_2O$ (Eigen cation), $H_5O_2^+$ (Zundel cation), $OH^-{\cdot}3H_2O$ (Eigen anion), and $H_3O_2^-$ (Zundel anion) [10]. All species are converted into each other by proton exchange and have different lifetimes. However, all of them contribute to the IR spectrum, showing high concentrations of short-lived ionic species [10]. Note that each act of proton exchange is accompanied by a charge transfer, followed by the adaptation of the surrounding molecules, or local polarization, which can be also seen in the absorption spectrum.

4.2.2 *The Spectral Signature of the Excess Proton*

The details discussed above assume the electrodynamic response of the excess proton in the dielectric spectrum, with a particular contribution to the infrared, terahertz, and microwave frequency ranges, down to radio waves. Note that the spectroscopic signature of the excess proton was identified in gas-phase water clusters only [21], and, its contribution to the spectrum of bulk water has not yet to be unambiguously identified. Nevertheless, by comparing the characteristic times of proton transport

[9]The transfer of the negative charge, represented by a hydroxyle ion is realized by analogy. The dynamics of the proton hole is realized by the proton transfer in the opposite direction.

[10]This property differentiates the modern models of water from those of a hundred years ago.

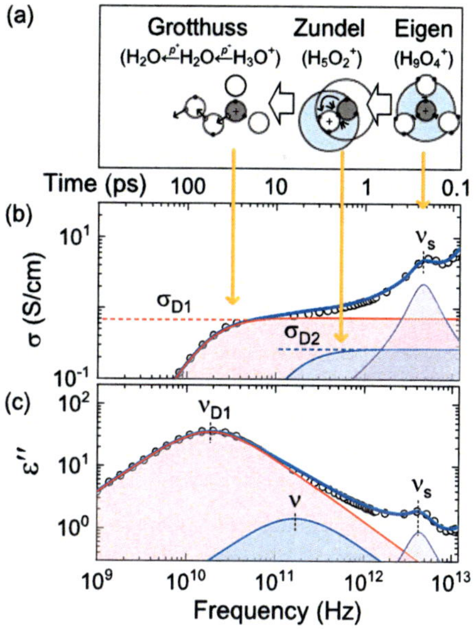

Fig. 4.2 The details of protonic transport in water on the picosecond-to-nanosecond timescale (**a**), compared to the corresponding frequency range, and the spectral features (color contours) of the spectra of (**b**) dynamical conductivity (σ) and (**c**) dielectric losses (ϵ'')

with the corresponding frequencies of the features of the spectrum, one can assign the dynamics of the excess proton to the characteristic parameters of the dielectric response of water.

Figure 4.2a shows the basic stages of the excess-proton dynamics: the interconversion of Eigen and Zundel cations by means of collisions and proton exchange, and the Grotthuss mechanism. These stages are depicted on a timescale which corresponds to the terahertz-to-gigahertz part of the dielectric response of water, shown separately on panels (b) and (c) of the figure. The parameters of the three main modes of this spectral region (two relaxors and one oscillator) are given in Table 4.1. There are two plateaus of conductivity, σ_{D1} and σ_{D2}, two characteristic frequencies of relaxations, ν_{D1} and ν_{D2}, and one central frequency of oscillation, ν_s, with a half-width, $\Delta\nu_s$. A comparison of the dielectric contributions $\Delta\epsilon$ of the relaxations show that the second relaxation is significantly smaller than the main (Debye) one.

The basic spectrally active stage is the Zundel cation, or the excess proton shared between two water molecules, whose intramolecular spectral activity is expected to be near the O–H stretching vibration $\nu_{1,3} \approx 3500\,\mathrm{cm}^{-1}$ (out of the scale in Fig. 4.2). Fundamental proton vibration has been observed by Dahms et al. [22] using 2D

Table 4.1 Parameters of the spectrum shown in Fig. 4.2

ν_{D1} (THz)	ν_{D2} (THz)	ν_s (THz)	$\Delta\nu_s$ (THz)	σ_{D1} (S/cm)	σ_{D2} (S/cm)	$\Delta\epsilon_{D1}$	$\Delta\epsilon_{D2}$
0.02	0.18	5.3	1.8	0.72	0.28	71	3.0

absorption spectroscopy in the form of a broad absorption band between 900 and 1500 cm^{-1}, with overtones assigned to the broad absorption continuum between 1,500 and 2,700 cm^{-1}. In addition, one should also expect the intermolecular spectral activity of the excess proton, which should appear in the terahertz frequency range. Proton dynamics are responsible for static conductivity too, thus, the dynamics of the solvated proton is expected to contribute to the whole frequency range from the IR down to low-frequency static conductivity σ_{dc}.

The essence of the proton current is the continuous mutual transformations of Eigen and Zundel cations (see Sect. 1.3). The Eigen cation is a hydronium ion solvated by three water molecules (see Fig. 4.2a), which has a spectral activity near the 5 THz oscillatory mode, ν_s, due to the vibration of the central ion in the shell of surrounding molecules (see Sect. 2.6.2). The Zundel cation is a transition point between two vibrational states. The switch between two Eigen states goes by proton tunneling, followed by the relaxation of the hydration shell (see the two circles around Zundel cation in Fig. 4.2a). The spontaneous tunneling of the charge induces the delayed adaptation of the nearest molecular dipoles, which presumably corresponds to the second relaxation (see the blue area in Fig. 4.2). As the polarization of the hydration shell is due to the reorientations of H_2O molecules, the second relaxation is also associated with molecular rotations. The local current induced by the migration of naked (unsolvated) ion contributes to the plateau σ_{D2}.

Continuing to increase the time of observation, we switch to a series of proton transfers, which contribute to another conductivity plateau σ_{D1}. This plateau can be considered as a current made by the dressed (solvated) ions, as the charge drags its solvation shell. That is why, in the real spectrum, this plateau is distorted by the solvation and the broadening of the peak ν_s. The lifetime of an ion in the oscillatory state between two transitions can be found from the half-width $\Delta\nu_s$ of this peak. Finally, a decrease of the conductivity below ν_{D1} is due to the finite lifetime of ionic species and due to their mutual electrostatic interactions, which, as discussed in Sect. 3.5, effectively reduce the cumulative mobility of the charge carriers at relatively long observation times (more than a few dozen picoseconds).

Thus, we correlated the intramolecular and intermolecular dynamics of the excess proton with the broadband dielectric spectrum of water and identified the characteristic frequency ranges, where we can expect the spectral signatures, which correspond to the vibrations, tunneling, translations, drift, and diffusion of excess protons and proton holes.

4.2.3 The Ionic (Protonic) Model of Water

Now, accounting for the details of proton transport described above, we formulate the ionic model of water. First of all, we represent water as an ensemble of H_2O molecules and intrinsic ions, H_3O^+ and OH^- (see Fig. 3.11), obeying thermal collisions, which transform to each other by the transfer of a proton (or proton hole). In other words, we admit spontaneous autoprotolysis and periodic proton exchange between water molecules and intrinsic ions, thus accounting for the nuclear (protonic) quantum effects. We assume that each molecule/ion obeys rapid oscillatory dynamics around the equilibrium position, being simultaneously in Brownian diffusion motion (see Sect. 4.2.1). The water is considered electrically neutral, containing an equal number of positive and negative ions (the electroneutrality principle), so that in equilibrium we have

$$\frac{n_w}{t_w} = \frac{n_\pm}{t_\pm}, \tag{4.1}$$

where n_w and $n_\pm$ are concentrations of H_2O molecules and ions, respectively. The concentration of H_2O molecules can be found as $n_w = n_0 - n_\pm \approx n_0$, where $n_0 =$ 55.5 mol/l is the full concentration of all species in water.

Figure 4.3 represents the details of atomic-molecular transport in water, relevant to its electrodynamic properties. For simplicity, the ions and H_2O molecules are represented by gray and white spheres, respectively, and the only one type of ions (H_3O^+) is considered. Protons and proton holes are depicted by small solid black and open circles, respectively, and oxygen atoms are assumed to be in the center of the each sphere. The numbers indicate the polarization relaxations that have a reflection in the spectrum.

Let us assume that at the initial moment, the H_3O^+ ion oscillates with a characteristic frequency ν_s inside the hydration shell (event 1 in the figure). This state we call a "dressed"-charge state. When the observation time increases, the excess proton (charge) escapes the hydration shell and becomes "naked." The transformation goes by the charge transfer between the molecular species. The excess proton changes the host molecule, being occasionally squeezed between two neighboring species (event 2). The charge hops the distance of 2.8 Å of two molecular radii, while the proton as a particle just sticks from one molecular species to another (i.e., does not hop at all). The time spent by the ion in the vibrational state determines the half-width $\Delta\nu_s$ of the ν_s mode. The subsequent relaxation of the naked state to the dressed state is characterized by the relaxation time $t_\pm$, coinciding with the secondary relaxation time $1/\nu_{D2}$. Due to continuous diffusion, the naked charge becomes dressed again at a distance l from the initial position, equal to $\sqrt{D_1 t_\pm}$, where D_1 is a diffusion coefficient of the solvated charge.

Further, the charge, separated from the parent molecule, diffuses independently, repeating a series of Zundel-to-Eigen cation transformations (event 3). The molecule abandoned by the charge moves like a neutral particle (event 4) until it again encounters another (or the same) charge and turns back to the ionic state (event 5). The time interval between the events of charge separation and ion reconstruction is the lifetime t_w of the molecule H_2O.[11] The time between events 1 and 5 is also the time of the main (Debye) dielectric relaxation [23]. Within the time t_w, an excess charge (the polarization configuration of molecules) passes a distance L in the laboratory frame of reference.

[11]Here, the lifetime is defined as the period between the transformations of a molecule into an ion by adding/removing one proton.

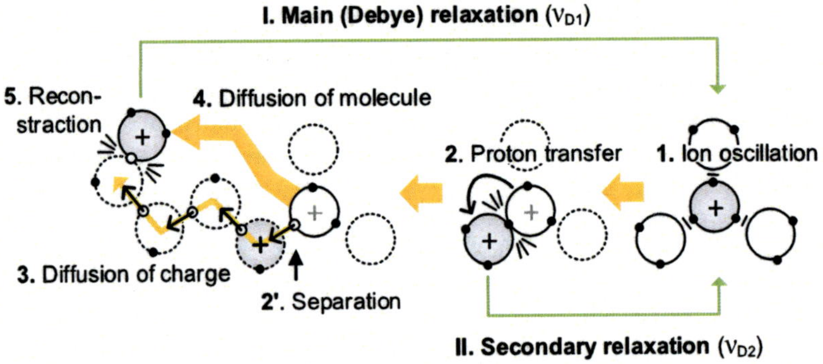

Fig. 4.3 The schematic of the atomic-molecular dynamics in water, which corresponds to the experimental dielectric spectrum shown in Fig. 4.2. Orange arrows show transitions between specific states (from right to left): (1) is the oscillatory dynamics of the excess proton (ion); (2) is the successful proton transfer from the ion to the nearest water molecule (charge transfer) followed by the adoption of the hydration shell; (3) is the series of excess charge transfers by the transformation of the states 1 and 2; (4) is the diffusion of a molecular species between two excess-proton states; and (5) is the reconstruction of a molecule and excess proton. Green arrows show two relaxation modes: (I) is the main (Debye) relaxation, and (II) is minor secondary relaxation (see text for details)

Comparing the processes in Fig. 4.3 and the spectral features in Fig. 4.2, we determine the following characteristic times: $t_s = \nu_s / \Delta\nu_s^2$ is the lifetime of an ion in the oscillatory state, $t_u = 1/\nu_{D2} - 1/\Delta\nu_s$ is the lifetime of an ion in the translational state, $t_w = 1/\nu_{D1}$ is the lifetime of a neutral H_2O molecule equal to the relaxation time t_{D1}, and $t_\pm = t_s + t_u$ is the lifetime of an ion. Table 4.2 contains the numerical values of these times at room temperature, calculated using experimental data from Table 4.1.

Further analysis requires an understanding of the structure of the diffusion mechanism, which is different for charges and molecules. Singwi and Sjölander showed [24] that in the case of two-component oscillatory-translatory motion, the diffusion coefficient is determined by

$$D_1 = l^2 / [6\,(t_s + t_u)]\,, \tag{4.2}$$

where D_1 is the effective diffusion coefficient of a species, l is the elementary diffusion step of oscillatory-translatory motion (see Fig. 1.25), and t_s and t_u are the lifetimes of a species in the oscillatory and diffusion states as defined above.

Figure 4.4 shows the spectrum of dynamic conductivity $\sigma(\nu)$. The spectrum has three conductivity plateaus: σ_{dc}, σ_{D1} and σ_{D2}, determined by the different mobilities of excess protons at different time intervals. In the short time interval (fractions of picoseconds) the ion oscillates at the bottom of the potential (see inset L in figure). The translational motion of ion goes through the short-range potential barrier (see inset R_2), which has a period of l. The corresponding naked-proton current manifests

Table 4.2 The parameters of the dynamical structure of water according to the ionic model

Parameter	Formula	Value at 298 K	Physical meaning
t_s (ps)	$\nu_s/\Delta\nu_s^2$	1.6	Lifetime of ion in oscillatory state
t_u (ps)	$1/\nu_{D2} - 1/\Delta\nu_s$	1.3	Lifetime of ion in translational state (regime of naked ion)
t_w (ps)	$1/\nu_{D1}$	49.5	Lifetime of H_2O molecule
$t_\pm$ (ps)	t_s+t_u	2.9	Lifetime of H_3O^+ and OH^- ions
$n_\pm$ (mol/l)	$n_0 \cdot t_\pm/t_w$	1	Concentration of H_3O^+ and OH^- ions (both short- and long-lived)
n_{pH} (mol/l)	$\sigma_{dc}/(C \cdot D_1)$[a]	10^{-7}	Concentration of long-lived H_3O^+ and OH^- ions only
D_2 (m^2/s)	$\sigma_{D2}/(C \cdot n_\pm)$	17×10^{-9}	Translational diffusion coefficient of the naked ion
D_1 (m^2/s)	$\sigma_{D1}/(C \cdot n_\pm)$	14.2×10^{-9}	Effective diffusion coefficient of the dressed ion
L (Å)	$\sqrt[3]{3 \cdot 0.74/(4\pi n_\pm)}$	8.4	Average distance between ions (separation-reconstruction diffusion length)
l (Å)	$\sqrt{6D_1 t_\pm}$	3.4	Elementary diffusion step of dressed ion
λ (Å)	$\sqrt{6D_2 t_u}$	2.9	Elementary diffusion step of naked ion

[a] $C = q^2/k_B T$

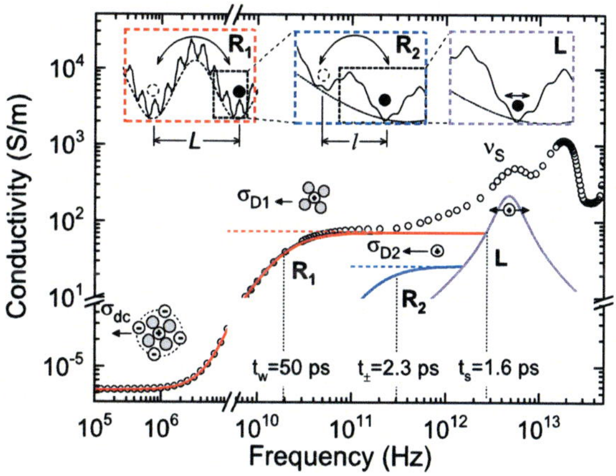

Fig. 4.4 The conductivity spectrum of water and its "structure" (color lines), corresponding to the different regimes of the diffusion of ionic species, depending on which periods of the complex potential, shown in the insets, are effective in the corresponding frequency range

itself as the conductivity plateau σ_{D2}. However, this plateau is hidden by the plateau σ_{D1}, which corresponds to the dynamics of hydrated (dressed) ion. This latter plateau lasts until the frequencies at which an interionic integration comes forward, and the conductivity decreases, showing the dielectric relaxation. The corresponding interaction potential (see inset R_1) reduces conductivity and the corresponding mobility even more (see Fig. 3.5 and the corresponding text).

The diffusion coefficient D of excess protons differs at each plateau (σ_{dc}, σ_{D1} or σ_{D2}) and is associated with the conductivity by the Nernst–Einstein relation:

$$\sigma_i = \frac{q^2}{k_B T} n_\pm D_i, \tag{4.3}$$

where subscript i means dc, $D1$ or $D2$, q is the elementary charge, k_B is the Boltzmann constant, $n_\pm$ is the concentration of charge carriers. The conductivity plateaus σ_{D1} and σ_{D2} are connected with the dielectric contributions $\Delta\epsilon_{D1}$ and $\Delta\epsilon_{D2}$ (see Table 4.1) of the corresponding relaxation bands ν_{D1} and ν_{D2} by the Debye formula:

$$\sigma_i = \varepsilon_0 \Delta\varepsilon_i 2\pi \nu_i. \tag{4.4}$$

Further, the average distance between the charges is determined by the concentration $n_\pm$:

$$L \approx \left(\frac{3}{4\pi n_\pm}\right)^{1/3}. \tag{4.5}$$

The system of (4.1)–(4.5), using experimental data from Table 4.1, allows one to determine the parameters of the atomic-molecular dynamics of H_2O molecules and H_3O^+ and OH^- ions in water. The parameters calculated at room temperature are given in Table 4.2, which constitutes the core parameters of the ionic model of water. The table shows that, within the ionic model, a microscopic description of the conductivity spectrum is achieved with the concentration of short-lived intrinsic ions of $n_\pm = 1$ mol/l on the average distance of 0.8 nm from each other. This concentration is 2% of the all the molecular species in water, but due to the ultrashort (2.3 ps at room temperature) lifetime, most of them disappear faster than they make a contribution to the static conductivity σ_{dc}. Nevertheless, any "snapshot" of water is represented by "swarms" of about 50 H_2O molecules per an H_3O^+ or OH^- ion.

Figure 4.5 shows the instantaneous water structure according to the above-calculated parameters of the ionic model (compare with Bernal–Fowler water in Fig. 1.5). It consists of electrostatically interacting molecular and ionic species. The electrostatic attraction of molecular dipoles by the ionic species increases the local density. The density between the ions of different signs is lower, as the forces on the molecular dipole, placed between the negative and positive charges, are balanced, and the center of the molecular dipole is in indifferent equilibrium. Note that the concept of water made of a linear combination of two components was introduced by Röntgen [25]. The idea of LDW and HDW was used for the explanation of water's thermodynamic properties [26], and recently verified by the neutron and X-ray diffraction techniques [27]. However, the structure of the areas of high and low density still lacks clarity on the microscopic level. The ionic model provides a microscopic picture, substantiating the existence of high- and low-density regions in water, and explaining the fast density fluctuations observed in scattering experiments by the dynamics of spontaneously appearing short-lived ionic species in sub-femtosecond timescales, with the collective relaxation effects lasting up to microseconds.

Thus, within the ionic model, the electrodynamics of water is determined by the thermal motion of short-lived ions and molecules. The main (Debye) dielectric relaxation is a result of the separation of H_3O^+ and OH^- ions (the polarization of the ionic atmosphere). The secondary relaxation is caused by the adjustment of the coordination sphere (solvation shell) of an ion following the spontaneous charge displacement. Static conductivity results from the migration of only long-lived ions, thus, is determined by the lifetime distribution of ionic species. The water structure shown in Fig. 4.5 can be considered as that made of swarms of hydrated ions with a concentration of about 2% of all molecular species. The boundary of the swarm determines the elementary intermolecular cell. The hydrated ion is thus a dynamic structural element of water according to the ionic model. The lifetime of an ion is limited to a few picoseconds. The ion, oscillating and drifting, after several collisions (on femtosecond timescales) transfers a charge (in the form of a proton) to a neutral H_2O molecule. In this way, the proton, always being a part of the ion, travels independently of the host molecule, while having the same diffusion coefficient. The electrodynamic response of water is determined by the interaction and interconversion of particles. The variety of the types of particle motion under conditions of mutual interaction, averaged over different observation times, is opposed to the variety of bonds, which are commonly used in the Bernal–Fowler model.

The ionic model differs from Bernal–Fowler water described in Chap. 1, because the former accounts for both short-lived and long-lived ions, while the latter operates with long-lived species only. Bernal–Fowler water can be considered as a special case of the ionic model, as it represents the diffusion-averaged D-structure, while the ionic model accounts for the instantaneous I-structure, the vibrationally averaged

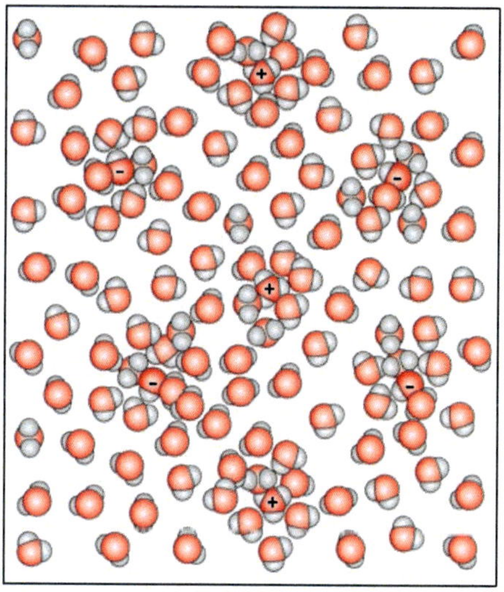

Fig. 4.5 The I-structure of water according to the ionic (protonic) model. Water consists of swarms of short-lived hydrated excess protons and proton holes in the form of H_3O^+ and OH^- ions (compare with Bernal–Fowler water in Fig. 1.5). The oxygen atoms are red, and the hydrogen atoms are gray. The ambipolar diffusion of excess proton and proton holes reproduce the experimental spectrum of dynamic conductivity (see Sect. 3.5)

V-structure, and the D-structure (see Sect. 1.1 for definitions) all together. The ionic model describes a wider range of experimental data. In particular it (a) eliminates the contradictions between the tracer diffusion (see Sect. 1.4), the mechanism of proton conduction and the nuclear magnetic resonance data; (b) uniformly describes the dielectric relaxation (microwave spectrum), the secondary relaxation (terahertz spectrum), and DC conductivity; (c) unambiguously determines the lifetimes of water molecules and ions; (d) qualitatively explains the thermodynamic and structural anomalies of water without the concept of hydrogen bonding; (e) provides a new insight to the modeling of the electrodynamic properties of water and ice at ultrashort (picosecond) timescales.

Thus, the molecular dynamics in water are complex, involving individual intermolecular proton migrations, and the collective behavior of short-lived ionic and molecular species. The excess protons, which are absent in the Bernal–Fowler model (see Fig. 1.7), create a spatial and temporal heterogeneity of water. The presence of short-lived ionic species was recently detected by means of infrared spectroscopy (see [10]). One can imagine the excess proton-proton-hole "gas", which exists in the frame of reference of the neutral water molecules, and moves in the potential landscape obeying Brownian motion in the electrostatic field of each other in the polarizable environment (see Fig. 4.5). The dynamics of the proton-hole gas quantitatively describes the broadband dielectric response of water and ice from DC current to terahertz (see Sect. 3.5).

4.3 Comparison of the Ionic Model with Other Microscopic Models of Water

Many different models of water have been suggested over the past hundred years [13, 28–35]. Most of them operate as a molecular system and are based on the ideas of Bernal and Fowler [13] discussed in Chap. 1. The ionic model of water described above is the natural extension of the Bernal–Fowler model but is not in line with the previous models. The ionic model does not postulate the molecular species as the basic structural element and it accounts for the previously missed nuclear quantum effects and processes at ultrashort (picosecond) time intervals, such as proton exchange and the formation of short-lived ionic species of high instantaneous concentration. To understand the range of the validity of this model, let us compare the ionic model with standard Bernal–Fowler water.

Standard Bernal–Forler water can be described as follows [36]. The structure of water (see Fig. 1.5) is viewed as having a uniform three-dimensional network formed by hydrogen-bonded molecules. The arrangement of this network is dynamically and structurally inhomogeneous. The water molecules continuously change their neighbors, so that the average lifetime of bonds is about a few picoseconds.

The ionic model describes water in a different way. Water is represented by an ensemble of neutral molecules and spontaneously formed short-lived intrinsic ionic species H_3O^+ and OH^-. Ions interact with each other and with neutral molecules. The spacetime heterogeneity of water is determined by the concentration of ions and the time of their relative conversion with neutral molecules by proton exchange at sub-femtosecond time intervals. The excess protons (and proton holes) can be considered as a proton-hole gas in a frame of reference water molecules.

Table 4.3 compares the ionic and Bernal–Fowler water models point by point. The Bernal–Fowler model is based on the assumption of the independent movement of its intrinsic ions, which hence have low concentrations and long lifetimes. The ionic model takes into account the interaction of ions, and hence they have a higher concentration and short lifetimes. The Bernal–Fowler model assumes a different interpretation of DC conductivity, microwave absorption, and the dielectric constant, while the ionic model considers them on the same basis. It is important to note that the ionic model also assumes the same parameters for water and ice, which is not the case for the Bernal–Fowler model. At the static limit, both the ionic and the Bernal–Fowler models describe the experimental data equally well, although they have different microscopic backgrounds.

There are several experimental facts, which go beyond the Bernal–Fowler representation of water, and can be explained in the frame of the ionic model only. These experiments are described below (more examples can be found in Chap. 5) and mainly concern the properties of water at ultrashort (picosecond) time periods, or the properties of water at the nanoscale, when the spatial-time heterogeneity of water comes to the fore, and cannot be neglected or averaged.

1. Optical Kerr effect (evidence for the spatial heterogeneity of water)
Taschin et al. showed [38] that the time-resolved optical Kerr effect allows one to observe the response of fast vibrational dynamics in water followed by a slower monotonic relaxation. The experimental spectra show characteristic features in the region of 50 and 200 cm^{-1}, which indicate the coexistence of two local configurations, which are interpreted as high-density water (HDW) and low-density water (LDW). The time interval, which corresponds to these frequencies, is considered in Sect. 2.6 and associated with the vibration dynamics of short-lived intrinsic ions of water and the relaxation of their hydration shell following the excess-proton transfer between ionic and molecular species. The spatial few-nanometer heterogeneity of water found by Taschin et al. is close to the average distance between the short-lived ionic species (see Fig. 4.5). The same period of heterogeneity was found by Huang et al. [39], who demonstrated the presence of density fluctuations in ambient water on a physical length-scale of about 1 nm by small-angle X-ray scattering.

2. Femtosecond mid-IR pump–probe spectroscopy (evidence for the time heterogeneity of water)
Woutersen et al. performed [40] a femtosecond mid-IR pump–probe study of the dynamics of the OH-stretching mode of HDO dissolved in D_2O. The orientational relaxation of the HDO molecules was observed to occur on either a very slow or a very fast timescale. Two discrete time constants of 0.7 and 13 ps were unambiguously identified. This result correlated with the previously mentioned study and assumes that two distinct molecular species exist in liquid water. The two discreet characteris-

Table 4.3 Comparison of the ionic (protonic) and Bernal–Fowler water models from the viewpoint of structure and electrodynamic parameters

	Bernal–Fowler model	Ionic (protonic) model
Interaction of intrinsic ions	No interaction	Ions electrostatically interact
Concentration of intrinsic ions	10^{-7} mol/l	1 mol/l (short-lived) and 10^{-7} mol/l (long-lived)
Dielectric constant	Described using polarizable molecular dipoles with fitting parameters, different for water and ice	Described by the relative displacement of positive and negative ions
DC conductivity	Described by the dynamics of independent H_3O^+ and OH^- ions	Both the DC conductivity and the dielectric relaxation are the result of the dynamics of interacting H_3O^+ and OH^- ions, whose mobility is frequency dependent
Microwave absorption and dielectric relaxation	Assumes rotational polarization mechanism	
Basic structural element	Molecule of H_2O	Molecules of H_2O and short-lived ionic species
The definition of pH	Autoionization	Thermal activation from the interaction potential
The main parameters, which are used for the description of the electrodynamic properties	The lifetime of bonds, the concentration of ionic defects, the concentration of bond defects, the percentage of unbound molecules, the molecular dipole moment, the molecular polarizability factor, bond energy. Total: 7	The lifetime of ions, the concentration of ions, the high-frequency mobility of ionic species, the low-frequency mobility of the species, the long-order interaction potential, the short-order interaction potential. Total: 6
Lifetime of the water molecule	11 h [37]	50 ps
Lifetime of ionic species	0.1 ms	2.9 ps

tic times assume the time heterogeneity of the dynamic water structure. Although the authors provide an explanation of the observed effect by the introduction of bounded and unbounded molecules, it is unclear why single-, double-, and triple-bounded molecules do not appear. The ionic model suggests an explanation in terms of two types of species, ionic and molecular, which appear in significantly different environments (see Fig. 4.5), and thus show two discreet characteristic relaxation times.

3. Fast mass and charge transfer in water confined in nanocapillaries (evidence for long-range order in water)
Bernal–Fowler water does not assume any deviation from the continual behavior of water down to a single-molecule dimension (0.3 nm). However, Holt et al. [41] found that already in the 2 nm pores (more than 6 molecular diameters), the water

permeability of the nanotube-based membranes was several orders of magnitude higher than those for continuum hydrodynamic models. Thus, the increased permeability is not a result of the absence of hydrogen bonds, but the result of the reduction of interaction on the intermolecular level, as is found between ionic species in the ionic model. The result of Holt et al. correlated with the experimental measurement of the electrical conductivity of nano-confined water by Artemov et al. [42], which was found to be five orders of magnitude higher than that for bulk water, and was explained by means of the ionic model of water. These results confirm that among short-order molecular–molecular interaction, there is a long-order intermolecular interaction in water, which can be changed by the confinement.

4. Water viscosity changes in a strong electric field (more evidence for long-range order in water)
Water is known to have anomalous properties being placed in a strong external electric field (several kV/cm or more) [37]. In particular, the phenomenon, known as a floating water bridge has been observed in the gap between two beakers under high voltage filled with chemically pure water [43]. Fuchs studied [44] this phenomenon and found that the bridge exhibits the "hidden" properties of water that only become visible in a strong electric field. In the Bernal–Fowler model, there are no such configurations of molecular dipoles, which are stable enough to form a free-hanging water wire no matter how large the external electric field is. On the contrary, the ionic model provides an intermolecularly polarizable environment, which exhibits long-order caused by the ionic "sub-lattice" (see Fig. 4.5). The relative displacement of the positive and negative intrinsic ions of water by an external electric field provides the electrostatic stress necessary to counteract gravity. For the analysis of the effect of the floating water bridge in the ionic model of water see Sect. 5.5.

5. The absence of hydrogen bonds between H_3O^+ ion and H_2O molecules (evidence for the long-range interaction of ions with water molecules)
Botti et al. studied [45] the microscopic structure of a concentrated HCl solution by neutron diffraction and found that the binding of water by the H_3O^+ ion occurs at a greater distance than that in the case of molecule–molecule interaction. Analyzing the scattering data, the authors found that the local density around ions is higher than that for neutral molecules as the intense peak at 1.45 Å appears on its radial distribution function. This shrinkage of water induced by ions is in line with the ionic model of water, which operates with swarms of water molecules around ionic species (see Fig. 4.5). The increase of the local density around ions should be compensated for by a decrease between the ions.

6. Semi-classical molecular-dynamic simulations (evidence that water is not an ensemble of long-lived H_2O molecules)
As discussed in Chap. 1, the density-functional theory (DFT)-based molecular dynamics are inapplicable for the modeling of the electrodynamics of water, as, minimizing the computational costs, it significantly underestimates the effect of proton exchange on water structure. The excess-proton trajectories for ion recombination predicted by DFT do not agree with the diffusive trajectories observed in

time-resolved spectroscopy, which are two orders of magnitude slower [46]. Bai and Herzfeld used [47] a force field for subatomic particles based on a semi-classical treatment of explicit valence electron pairs with heuristic potentials trained on the structural and thermodynamic properties of water monomers and dimers in all of their protonational states. The authors found that about 1 mol/l of simulated water molecules have an anomalously low O–O distance of less than 2.5 Å, which assumes no barrier for proton transfer and indicates the formation of ionic pairs. This concentration of ionic pairs is in remarkable agreement with the concentration of short-lived species in the ionic model. The authors also observed no spontaneous ionization in over 10 ns, which means that ion pairs annihilate in sub-nanosecond timescales. This is also consistent with the ionic model, which assumes a picosecond lifetime of ionic species. The special role of the ultrashort interactions, and the formation of the corresponding short-lived ionic species, is consistent with the elusive rise of the dissociation constant K_w of water with increasing pressure. Thus, the results by Bai and Herzfeld provide more satisfactory agreement with experiment than DFT does with different functionals and indirectly support the ionic model.

Summarizing, the comparison of the hundred-year-old Bernal–Fowler model with the ionic model shows the tremendous advantages of the latter in the description of ultrashort and nanoscale dynamics, and the corresponding experimental observations. The experiments of the past decade reveal the significant spatial-time heterogeneity of water, which can be neglected at large scales and relatively long observation times, but which cannot at picosecond or at nanometer scales. As Bernal–Fowler water does not account for the heterogeneity of water and nuclear quantum effects, the experimental data described above can be explained by the ionic model only. Thus, the ionic model has wider application and validity range and allows one to account for more dynamic properties of water than models based on the concept of a hydrogen-bound network of equivalent molecular species.

4.4 Instantaneous Structure of Water and Ice

Figure 4.6 shows the I-structure of water and ice according to the ionic model. The inset shows the effective potential for the central ion (excess proton). The potential has two characteristic periods and differs for water and ice. The height of the potential peaks and the length of periods are in accordance with the barriers (activation energies) of conductivity (see Table 3.2) and the diffusion lengths shown in Table 4.2. The spatial period of a smaller (by period) potential associated with diffusion length $l \approx$ 3.4 Å, which corresponds to the elementary step of the translational diffusion of the ion. The period of this potential is approximately the same for water and ice, but the amplitude differs by about 0.4 eV. The larger-by-period potential (see the enveloping dashed black line) associated with the rearrangement of the ionic atmosphere of the central ion, and associated with the second characteristic diffusion step $L \approx 0.84$ nm, which is half the unit step of DC conductivity. The height of the enveloping potential is determined by the difference between the activation energies of σ_{dc} and σ_{D1} (see

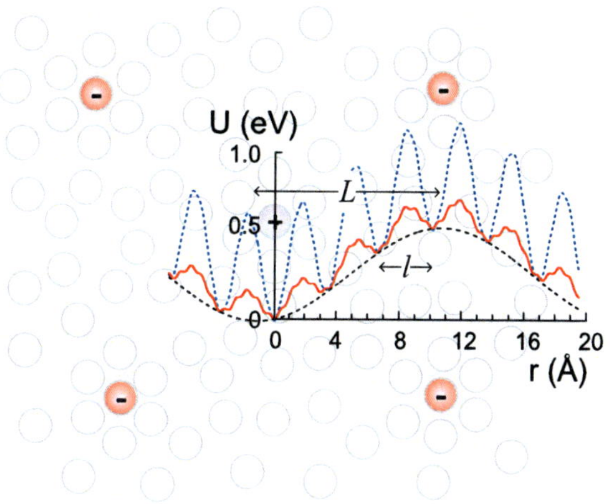

Fig. 4.6 The I-structure of water and ice according to the ionic model. Open circles are neutral H_2O molecules, colored circles are H_3O^+ and OH^- ions. The latter act as modulators of electrostatic potential. The inset shows the electrostatic potential U(r) for the central excess proton in the form of H_3O^+ for water (red) and ice (blue). The black dotted line is the potential enveloping curve, the same for water and ice

Table 3.2), and for water and ice is 0.2 eV. Thus, the main difference between water and ice from the electrodynamic point of view is the height of the barrier of the charge diffusion, which is determined by the probability of proton transfer. In other words, the I-structure of water and ice is the same, but the time constants are different. The excess protons in ice need more time to overcome the potential than those in water. The period of spatial heterogeneity of ice and water is the same, while the period of the temporal heterogeneity is different. Note that in [39] the regions of LDW and HDW with a characteristic size of 1 nm have been identified by neutron scattering. This length is close to the period L of the larger potential shown in Fig 4.6.

Figure 4.7a–d shows the I-structure of water in different representations. Part (a) shows the molecular structure, where colored circles represent the short-lived ionic species, and the white circles are the neutral water molecules. Part (b) shows the intermolecular network of electrostatically interacting ions. The field lines between the ionic species prescribe the preferential direction of the alignment of the molecular dipoles, and also depict the long-order. Although the field lines fluctuate on picosecond timescales, following the displacement of the charges, the ionic species form the intermolecular sub-lattice, which is independent of the molecular sub-lattice. The competition between the long-order ionic and short-order molecular lattices is responsible for the transition between water and ice.

Table 4.4 shows the structural and dynamic parameters of water and ice. The numerical values were calculated on the basis of spectroscopic data using the formulas discussed in Sect. 4.2.3. Data provided at 0 °C, where ice and water can exist at thermodynamic equilibrium. The instantaneous concentration of ionic species and the corresponding spatial parameters are close for ice and water. However, the time parameters differ significantly. In particular, the lifetimes of ionic and molecular species, $t_{\pm}$ and t_w, change by six orders of magnitude, following the corresponding shift of the dielectric relaxation time (see Fig. 3.9). Thus, although the I-structures of

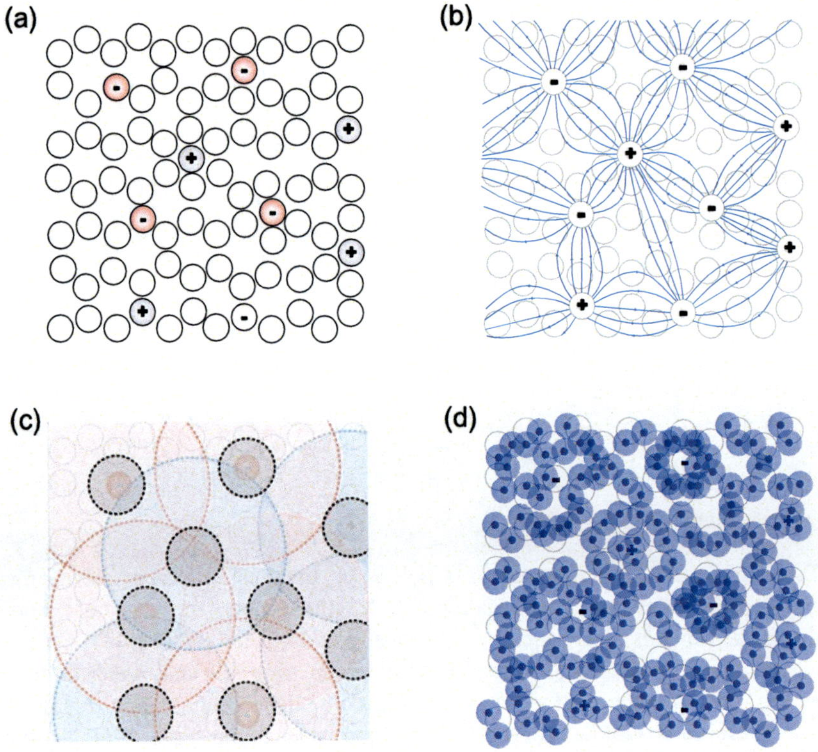

Fig. 4.7 The different representations of the same I-structure of water and ice: **a** the molecular structure (colored circles represent spontaneously born short-lived ionic species); **b** the electric field lines between excess charges; **c** polarization spheres around ions (small for the hydration shell and large for the ionic atmosphere); and **d** the protonic density map

Table 4.4 The structural and dynamic parameters of water and ice at 0 °C according to the ionic model. For the meaning of variables see text and Table 4.2

	t_w	$t_\pm$	$n_\pm$ (mol/l)	$D_\pm$ (m^2/s)	L (Å)	l (Å)
Water (0 °C)	120 ps	2.1 ps	0.95	$1.0 \cdot 10^{-8}$	6.8	7.0
Ice (0 °C)	130 μs	2.3 μs	0.87	$1.1 \cdot 10^{-14}$	3.5	3.7

ice and water are the same, they have quite different diffusion-averaged D-structures. And, as the formation of the ionic sub-lattice (see Fig. 4.7b) is determined by the stability of ionic species, this sub-lattice is more stable in ice, where the lifetime of ionic species, $t_\pm$, exceeds the characteristic time of molecular thermal fluctuations by many orders of magnitude. In other words, the molecular lattice "wins" the competition with the ionic lattice in liquid water, quickly destroying any long-order structures, while in ice the ionic sub-lattice dominates.

Part (c) in Fig. 4.7 shows the two characteristic polarization spheres formed by the ionic and molecular species. The small-gray spheres are the hydration shells of ions, which are formed by the nearest molecular coordination sphere of ionic species. As discussed in Sect. 4.2.2, the time lag between the displacement of the central ion and the adaptation of its hydration shell is responsible for the secondary relaxation of water. This relaxation is not observed in ice, because the lifetime of ionic species $t_{\pm}$ significantly exceeds the relaxation time of the hydration shell. The large-colored (red or blue) spheres represent the ionic atmospheres, which are formed by the ions of the opposite sign around the central ion. As one can see, these spheres overlap, and have a radius equal to the average distance between ionic species L. The polarization of these large spheres is responsible for the main (Debye) dielectric relaxation of both ice and water (see Sect. 3.5 for the quantitative analysis).

Finally, part (d) in Fig 4.7 shows the instantaneous proton density. Each hydrogen atom of a water molecule is shown by the black dot, surrounded by a blue area. The latter represents the diameter of the proton delocalization. Each overlap of the proton delocalization area with the nearest molecule is a potential "bridge" for proton transfer, which can be interpreted as a manifold of the complicated concept of hydrogen bonding (see Sect. 4.1 for details). One can see that the proton density is higher around the ionic species. The sub-picosecond thermal fluctuations of the proton density in water are, in principle, observable by neutron scattering.

Thus, the electrodynamic properties of ice and water allow one to reconstruct the I-structure, whose dynamics gives the slower V- and D-structures. Interestingly, a single parameter, the height of the potential barrier for proton transfer, is responsible for the global transformation of the dielectric spectra of ice and water. This barrier height is 3.5 times larger for ice than that for water (see Fig. 4.6). In other words, water and ice have an identical I-structure, and their D-structure differs only by the potential barrier of charge transfer. For water, in particular, the barrier height is 0.2 eV (20 kJ/mol) [48]. Protons randomly walk by hopping over neutral H_2O molecules to form short-lived H_3O^+ OH^- ion pairs.

The separated charges occur in water and ice in high concentrations, $n_{\pm}$, which accounts for the high value of the static dielectric constant, $\epsilon(0)$. However, the equilibrium of the separated charges is dynamic. It is maintained by two competitive mechanisms, the self-dissociation of H_2O molecules and the recombination of the $H_3O^+OH^-$ ion pairs. A decrease in temperature slows down proton diffusion, thus shifting the equilibrium to increase the lifetime of the separated charges. This slightly increases both, $n_{\pm}$ and $\epsilon(0)$, while the general temperature dependence of $n_{\pm}$ and $\epsilon(0)$ is preserved at the water-ice phase transition. The difference between the activation energies of proton diffusion in water and ice determines the latent heat of the water-ice phase transition.

4.5 The Microscopic Origin of the Electrodynamic Properties of Water and Ice

4.5.1 Dielectric Relaxation and DC Conductivity

In the ionic model, the dielectric (Debye) relaxation and the DC conductivity have the same origin: the ambipolar diffusion [49] of intrinsic ions. This type of diffusion, in which charges of two signs interact with each other is responsible, in water, for both the microwave conductivity σ_{ac}, and the static conduction σ_{dc}. Note that in the Bernal–Fowler model σ_{ac} and σ_{dc} are independent (see Sects. 1.3 and 2.3.2). Figure 4.8 illustrates the difference between the two models, showing three types of charge diffusion in different environments. Part (a) shows the spectrum of independent charged particles. The ionic conductivity σ_i is determined by the concentration of ions $n_{\pm}$ by

$$\sigma_i = qn_{\pm}\mu_i, \tag{4.6}$$

where q is the charge, and μ_i is the particle mobility. If there are no external forces acting on the conducting particle, the conductivity is frequency-independent for all frequencies below the collision frequency. However, ions in the condensed medium strongly interact with the surrounding species.

Part (b) shows the case in which the charged ion is moving in the periodic potential formed by the crystalline lattice. The dynamics of the selected ion can be split into

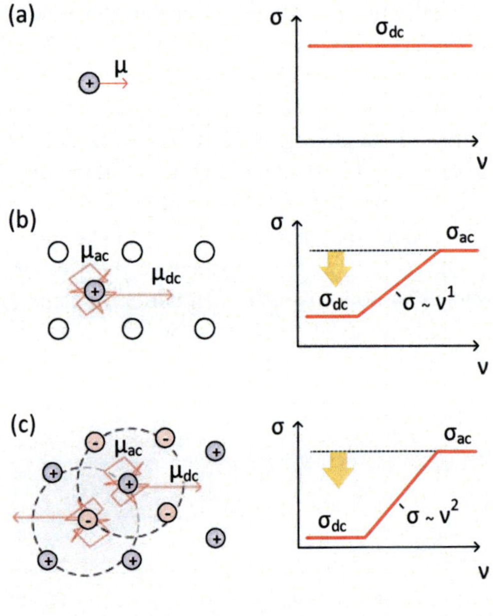

Fig. 4.8 Three types of motion of the ionic species in dielectric medium: **a** "free" motion; **b** stationary lattice; **c** ambipolar diffusion. The left columns show the microscopic picture (μ is mobility), the right column shows the corresponding conductivity σ spectrum

two fragments. The first fragment is "in-cage" dynamics with the mobility μ_{ac}, which manifests at high frequencies. The corresponding AC conductivity σ_{ac} is equivalent to that observed in part (a). The second fragment is the low-frequency dynamics with effective mobility, μ_{dc}, reduced by the interaction of the ionic species with the stationary lattice. The corresponding low-frequency conductivity σ_{dc} is lower than the high-frequency conductivity.

Finally, part (c) represents the case of water, in which the ionic species obey ambipolar diffusion. As in the previous case, there are two characteristic mobilities. The high-frequency mobility μ_{ac} corresponds to the dynamics inside the ionic atmosphere (see dashed circles), while the low-frequency mobility, μ_{dc}, is a result of the dynamics of the same particle slowed-down by the ionic atmosphere. Note that the transition regions between the high- and the low-frequency plateaus in parts (b) and (c) are different. For the stationary lattice the conductivity $\sigma(\nu)$ is proportional to ν^1 (see Fig. 4.8), while ambipolar diffusion gives $\sigma(\nu) \sim \nu^2$. The latter is exactly the same as for the Debye relaxation (see Sect. 2.3).

Ambipolar diffusion is described by the system of equations:

$$\begin{cases} m\gamma_1\dot{x} = \kappa(x - X) - f_1(t), \\ M\gamma_2\dot{X} = \kappa(X - x) - f_2(t), \end{cases} \tag{4.7}$$

where m and M, x and X are the masses and coordinates of the central ion and its ionic atmosphere, respectively, γ is a damping, and f_1 and f_2 are stochastic forces. The first term on the right-hand side represents the electrostatic coupling with the elastic spring constant κ, which connects the ion with the center of its ionic atmosphere, both of which obey Brownian motion. System (4.7) corresponds to the simplified case considered in Sect. 3.5.

The dynamic conductivity in the whole frequency range can be found from 4.7 by the Fourier transform of the velocity–velocity correlation function:

$$\sigma(\omega) = \frac{q^2 n_\pm}{kT} \int_0^\infty \langle \dot{x}(0)\dot{x}(t)\rangle e^{-i\omega t} dt, \tag{4.8}$$

which gives for the frequency-dependent dynamic conductivity:

$$\sigma(\omega) = \frac{\sigma_{\rm dc} + \omega^2\tau_{D1}^2\sigma_{ac}}{1 + \omega^2\tau_{D1}^2}, \tag{4.9}$$

where $\sigma_{dc} = qn_\pm\mu_{dc}$, $\sigma_{ac} = qn_\pm\mu_{ac}$, and the relaxation time τ_{D1} is

$$\tau_{D1} = \frac{\gamma_1\gamma_2 mM}{(m\gamma_1 + M\gamma_2)\kappa}, \tag{4.10}$$

and the low- and high-frequency mobilities are

$$\mu_{dc} = \frac{1}{k_B T} \frac{(f_1^2 + f_2^2)\tau_{D1}}{(m\gamma_1 + M\gamma_2)^2}, \tag{4.11}$$

$$\mu_{ac} = \frac{1}{k_B T} \frac{f_1^2 \tau_{D1}}{m^2 \gamma_1^2}. \tag{4.12}$$

Equation 4.9 is similar to the Debye formula for dielectric relaxation, but it additionally includes the static DC conductivity, σ_{dc}, which is missing in the Debye model of the orientational polarization of permanent dipolar species. Thus, the ionic model has a wider range of validity than the Bernal–Fowler model, allowing one to put both the high- and the low-frequency conductivity on the same footing.

Figure 4.9 shows the best fit of the function given by (4.9) to the experimental data with parameters from Table 2.2. One can see that the experimental data are comprehensively described by the model up to 0.1 THz.

Thus, the diffusion of intrinsic ions in water allows one to describe both static conductivity and Debye relaxation on the same basis: the diffusion of ions in the field of a central force averaged over different observation times. The experimental conductivity spectrum of water is described without using the model of orientational polarization. Instead the bipolar diffusion imposed by the Coulomb interaction of ions allows one to connect the low-frequency DC conductivity of water with the dielectric (Debye) relaxation.

Assuming in (4.9)–(4.12) that $\gamma_2 \gg \gamma_1$ means that the friction for the ionic atmosphere is much higher that that for the central ion, for the relaxation time, τ_{D1}, and DC conductivity, σ_{dc}, we obtain

$$\tau_{D1} = \frac{\gamma_1 m}{\kappa} = \frac{1}{\mu_{ac}\kappa}, \tag{4.13}$$

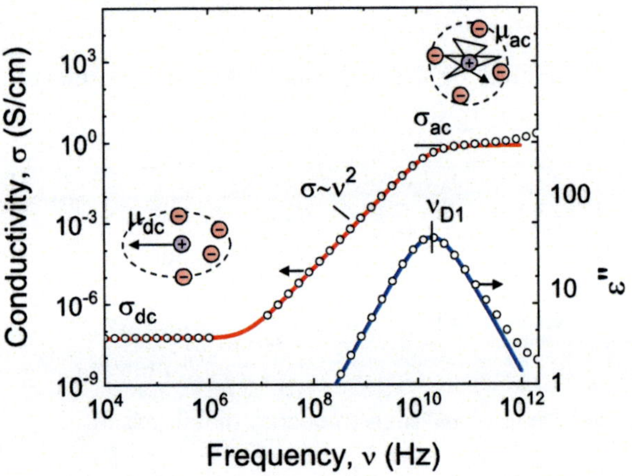

Fig. 4.9 The spectra of dynamical conductivity, σ, and the imaginary part of the dielectric constant, $\epsilon'' = \sigma/\epsilon_0/(2\pi\nu)$. The color lines are fit according to the model (4.9). Insets show two limits of the diffusion of the same charge: the high-frequency "in-cage" dynamics with mobility μ_{ac}, and the low-frequency long-order dynamics with effective mobility μ_{dc}

and

$$\sigma_{dc} = \frac{q^2}{k_B T} n_{\pm} \frac{\left(f_1^2 + f_2^2\right)}{M^2 \gamma_2^2} \tau_{D1}. \tag{4.14}$$

Formula 4.13 connects the dielectric relaxation time τ_{D1} with the elastic interaction constant, κ, thereby defining its physical meaning as the coupling time of the charge with the center of its ionic atmosphere. Formula 4.14 shows that the low-frequency DC conductivity, σ_{dc}, is a result of the ambipolar diffusion of intrinsic ions. As the friction constant γ_2 of the ionic atmosphere is in the denominator, the lower the constant, the higher the static conductivity. In Chap. 5 (see Sect. 5.2) we discuss the walls of the confining matrix that can effectively reduce the coupling constant, which leads to the increase in the static DC conductivity of confined water by up to five orders of the frequency magnitude, depending on the confinement diameter.

Taking $\gamma_1 = 4.3$ THz (see Table 3.5) and the mass of the hydrated ion $m \approx 4 \cdot m_{H_2O} = 1.5 \cdot 10^{-25}$ kg, from (4.13), we obtain $\kappa = 7 \cdot 10^{-2}$ N/m, which is close to the surface tension of water $\sigma_s = 7.8 \cdot 10^{-2}$ N/m [50]. In addition, the value of κ is in good agreement with that obtained in the experiments on the cavitation of water. The experimental value for the critical pressure of water near the cavitation threshold is $p_{cr} = 28$ MPa [51]. Assuming that the size of the discontinuity region is the distance between ionic species $L = 0.8$ nm (see Table 4.2), we obtain $\kappa_{exp} = p_{cr}/L = 4 \cdot 10^{-2}$ N/m. Although this value is half the friction caused by the ion–ion interaction, it is still of the same order of magnitude. Consequently, the tensile strength and the surface tension of water can be associated (at least partially) with interaction among its intrinsic ionic species.

4.5.2 How Microwave Ovens Work

Microwave ovens utilize the interaction of electromagnetic waves (the same as radio waves but with a higher frequency) with liquid or solid media, heating it up. The common source of the electromagnetic waves of the microwave region is a magnetron,[12] the device which generates electromagnetic waves accelerating/decelerating a beam of electrons using a magnetic field. When the microwaves[13] contact the medium, the oscillating electric field of the wave causes the internal charges of the medium to move by the Lorentz force. These microscopic dynamics cause the rise in temperature, allowing the operator to heat objects or cook food.

[12]There are also backward-wave oscillator, traveling-wave tube, klystron, and other sources of high-frequency electromagnetic waves.

[13]In fact, the working frequency of the microwave oven is around a few GHz, and the corresponding wavelength is about 10 cm in air, and a few cm in water. Thus, the word "microwaves" is just used historically to differentiate them from radio waves and does not reflect the true wavelength of the corresponding electromagnetic wave, which is in fact several orders of magnitude greater.

The amount of power, P, (in W/m^3) that is absorbed by the medium is determined by the level of absorption of electromagnetic waves by the medium (i.e., heating) and is given by

$$P = \varepsilon''(\omega)\varepsilon_0\omega E^2 = \sigma(\omega)E^2, \tag{4.15}$$

where $\omega = 2\pi\nu$ is the angular frequency, $\sigma(\omega)$ is the dynamic conductivity, ϵ_0 is the permittivity of vacuum, and E is the potential gradient (strength of the electric field in V/m), induced by the electromagnetic wave. Equation 4.15 means that the conductivity is a measure of dielectric losses. The higher the conductivity, the more effective the absorption of electromagnetic waves.

The typical operating frequency of the domestic microwave oven is 2.45 GHz. Microwaves are especially good for heating water, because it has a high level of absorption (the high dynamic conductivity, $\sigma(\omega)$) in the corresponding frequency range. Although there is a special defrosting regime in domestic microwave ovens, ice is quite transparent for microwaves, thus, does not interact with electromagnetic waves of the corresponding frequency as effectively as water. However, the ice surface is always covered with a quasi-liquid layer, which effectively absorbs microwaves, and then transfers the heat to the rest of the ice. Thus the defrosting regime is less powerful than the normal heating regime to avoid the evaporation of the quasi-liquid layer prior to heat transfer to the ice volume. Figure 4.10a shows the dielectric spectra of water at room temperature and at 100 °C. The operating frequency of the domestic microwave oven is shown by the arrow. It lies on the left-hand side of the Debye relaxation band (see blue bell-shaped contour) and does not correspond to the maximum adsorption of electromagnetic waves. However, the electromagnetic wave sources of the higher frequencies are much more expensive, and the usage of this frequency is not economically viable.

A common misconception is that the microwaves rotate the molecular dipoles of water, allowing them to move faster and in this way increasing the temperature of the medium. In fact, the characteristic frequencies of the molecular orientations (both free and hindered) are much higher than those used in microwave ovens, and lie in the terahertz frequency range (see Sect. 2.6), thus, cannot be directly responsible for microwave heating. We showed above that the dielectric relaxation (and therefore the microwave absorption) is caused by the dynamics of short-lived ionic species of H_3O^+ and OH^-, most of which live about 3 ps, although some of them exist significantly longer, contributing to the static conductivity. The ambipolar diffusion of interacting ionic species was shown to be responsible for the dielectric response of water below 0.1 THz (see Sect. 4.5.1), including the operating range of microwave ovens. Thus, according to the ionic model, the short-lived ionic species are dragged by the field, heating the environment as shown in Fig. 4.10b. As the ionic species are surrounded by neutral molecules, they are also dragged, but the main drivers of heating are ions.

Taking $E = 100$ V/m (the maximal electric field strength in the domestic microwave oven), $n_\pm \approx 1$ mol/l, $\kappa = 0.07$ N/m, and using (4.9), (4.12), and (4.13), we get from (4.15) that $P \approx 10$ W/cm^3. Thus, to heat a cup of water (250 ml) from

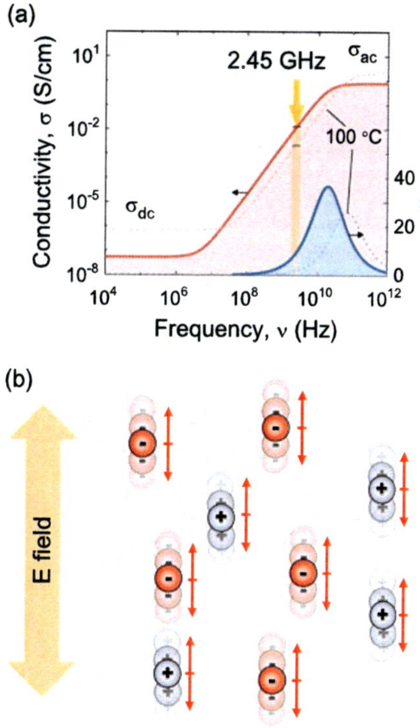

Fig. 4.10 On the mechanism of the absorption of microwaves by water. **a** The spectra of dynamic conductivity, σ, and the imaginary part of the dielectric constant, ϵ'' at 25 °C (solid) and 100 °C (dashed). The arrow shows the standard microwave oven frequency. **b** The microscopic mechanism of the absorption of microwaves. The external electric field drags the intrinsic ions of water, whose relative displacement results in the appearance of effective intermolecular dipole moments, whose direction follows the direction of the field. The neutral water molecules are depicted by white circles

25 to 100 °C, we need the time $t = c\Delta T/P$ (about 2 minutes), where $c = 4.2$ kJ/kg · K is the heat capacity of water. Note that the relaxation band changes with temperature, as shown in Fig. 4.10a. Thus, the effectiveness of the heating changes with time, as the conductivity level at the working frequency decreases by almost an order of magnitude with the increase of the temperature (compare small black sticks on the graph); the hotter the water gets, the harder it is to heat further.

Interestingly, the microwave part of the spectrum of water is little affected by impurities, such as foreign solute (see Sect. 5.1). That is why, the salt water (or soup) heats almost as quickly as fresh (distilled) water. Another case is confined water, whose spectrum is discussed in Sect. 5.2. While the static conductivity of confined water is significantly higher than that for the bulk water [42], the high-frequency conductivity, σ_{ac}, is the same, or lower, than in bulk water. Thus, the effectiveness of heating confined water is lower in comparison with the bulk.[14] However, when the pore size reaches a few nanometers (a very rare case which can be found in artificial materials), the static dielectric constant, $\epsilon(0)$, of confined water drops down to the value of 2 [52]. As the static dielectric constant is connected to the dielectric

[14]Note that microwaves cannot penetrate deeply, as they effectively absorb by water, which attenuates the strength of the electric field. Thus, a large volume of water unavoidably heats from the boundaries rather than in the volume. However, in food, water usually presented is relatively small volumes, not in one piece, which increases the efficiency of microwave heating.

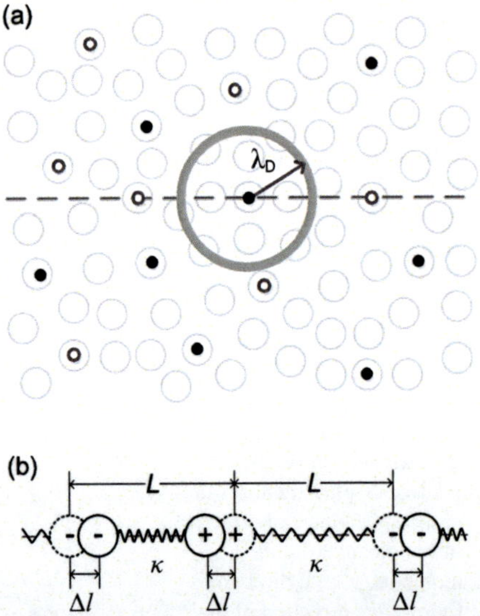

Fig. 4.11 **a** The "gas" (or "plasma") of excess protons (black dots) and proton holes (open dots) in the frame of reference of neutral molecules. The gray circle shows the Debye screening radius. **b** The chain of charges, showing the relative displacement Δl of charges in the presence of electric field. The initial (unperturbed) position of charges is shown by the dashed circles. The springs with a constant κ represent the electrostatic interaction

relaxation, its reduction means the reduction of the absorption of microwaves. Thus, if one believes in the value obtained in [52], we should expect that water which is confined in channels a few nanometers wide will not heat in a microwave oven. In other words, the water confined in nanometer pores will be totally transparent to microwaves.

4.5.3 The Dielectric Constant

The excess protons and proton holes in water can be considered as a "gas" of charged particles or a protonic plasma, in which the charges exist in the frame of reference of neutral water molecules. Such a two-component molecular system is schematically shown in Fig. 4.11a. The neutral molecules (large white circles) form the thermally fluctuating background, and the charged excess protons (black dots) and proton holes (open dots) form another subsystem, which exist in the frame of reference of the first system, but also electrostatically interact. The Debye screening length (gray circle) is determined by the concentration $n_{\pm}$ of protons and proton holes:

$$\lambda_D = \left(\frac{q^2 n_{\pm} N_A}{\varepsilon \varepsilon_0 k_B T} \right)^{-1/2}, \tag{4.16}$$

where N_A is the Avogadro number. Equation 4.16 for the concentration $n_{\pm} = 1\,\text{mol/l}$ (see Table 4.2) gives $\lambda_D = 0.5\,\text{nm}$. This Debye length means that our plasma of protons and holes can be considered quasi-neutral on scales grater than $2\lambda_D = 1\,\text{nm}$.

Figure 4.11b shows the 1D chain of electrostatically interacting charges located along the dashed line in Fig. 4.11a. The average distance L between the ions is roughly determined by their concentration:

$$L \approx (n_{\pm})^{-1/3}, \tag{4.17}$$

and is approximately equal to 1 nm. The alternating external electric field causes the displacement of the charges by Δl, whose amplitude value $l = 3.4$ Å is determined by the lifetime of the ionic species (see Table 4.2). The restoring force, which acts on displaced ions, is $F = \kappa \Delta l$. According to (4.13), $\kappa = \gamma_1 \mathrm{m}/\tau_{D1}$, where τ_{D1} is the temperature-dependent dielectric relaxation time.

The dielectric constant of ionic plasma can be found in [49] (see also (3.24) and the text below it):

$$\varepsilon(0) = \varepsilon_{THz} + \frac{\Omega_p^2}{\omega_0^2}, \tag{4.18}$$

where

$$\Omega_p = \left(\frac{n_{\pm} q^2}{m\varepsilon_0}\right)^{1/2} \tag{4.19}$$

is the plasma frequency equal to 1.8 THz, and $\omega_0^2 = \kappa/\mathrm{m}$ (see Table 3.5). Equation 4.18 gives $\epsilon(0) = 79$ for water.

Thus, within the framework of the ionic model, the high dielectric constant of water, $\epsilon(0) \approx 80$, is determined by the high concentration of its spontaneously formed intrinsic ionic species, whose relative displacement creates a dipole moment greater than the dipole moment of an individual molecule. The next section shows that the relative contribution ϵ_{THz} of the orientation of molecular dipoles to the static dielectric constant is only about 5%.

Note that there are other reasonable models for the dielectric constant, including the famous Clausius–Mossotti, Debye–Onsager, and Kirkwood–Fröhlich models. Although these approaches provide the correct values of the dielectric constant of polar liquids [53], they still lacking clarity on the microscopic level (see, for instance, the detailed analysis by Hippel [54]). The basic idea of these models is the concept of the local field (see Sect. 2.4.2). Such an approach leads to the additional polarizability of individual H_2O molecules represented by the emperical Kirkwood g-factor.[15] However, the experimental facts are against local polarizability (see Onsager's papers and the polarizability catastrophe [56]). In addition, classic neutron diffraction shows no change of the H_2O molecule dipole moment in comparison with

[15]In his classic 1939 paper, Kirkwood linked the macroscopic dielectric constant of polar liquids to the local orientational order as measured by the g-factor (later named after him) and suggested that the corresponding dielectric constant at short-range is effectively equal to the macroscopic value just beyond the distance of molecular magnitude [55].

that in the gas [57]. IR peaks of the intramolecular dynamics of water are also stable, showing no shift between vapor and liquid. The shifts 3,756 → 3,500 cm^{-1} (7%) and 3,652 → 3,600 cm^{-1} (< 2%) [58] are too small to be responsible for the high molecular polarizability required for dielectric constant interpretation in terms of the local field. Another intermolecular polarization mechanism is required. A possible mechanism was described above, and it finds an unambiguous interpretation in the ionic model of water.

4.5.4 The Second Relaxation and Terahertz Spectrum

At short time intervals, we can neglect the dynamics of the ionic atmosphere and write a single equation of motion for the central ion performing harmonic oscillations in a soft parabolic potential:

$$m\ddot{x} + m\gamma\dot{x} + m\omega_0^2 x = qE_0 \exp(-i\omega t), \tag{4.20}$$

where ω_0 is the eigenfrequency of the oscillations, and E_0 is the amplitude of the periodic external electric field. Figure 4.12a shows the equivalent mechanical scheme for (4.20). The scheme and the corresponding equation coincide with the forced oscillations with friction, considered in [59], and is similar to the dynamics of the Kapitza pendulum [60, 61] in which the mathematical pendulum is replaced by a spring oscillator. Introducing $x_0(t) = qE(t)/m\omega_0^2$, which represents the change of the equilibrium coordinate of the oscillations by the external field $E(t) = E_0\exp(-i\omega t)$, we get a simple equation of a damped harmonic oscillator with a moving equilibrium point:

$$m\ddot{x} + m\gamma\dot{x} + m\omega_0^2 (x - x_0(t)) = 0, \tag{4.21}$$

where to let the particle conduct, we assume the friction has the following form: $\gamma(t) = \gamma_0\exp(-t/\tau_r)$, where τ_r is the length of time an ion is in the oscillatory state.

In terms of the dynamic conductivity, $\sigma(\omega) = n_\pm q\dot{x}(\omega)/E(\omega)$, (4.21) has the following solution [59]:

$$\sigma(\omega) = \frac{n_\pm q^2 \cdot \gamma_0(1 - i\omega\tau_r)}{m\left[\gamma_0\tau_r(\omega_0^2 - \omega^2) + \gamma_0^2 - \omega_0^2 - i\omega\gamma_0(1 + \gamma_0\tau_r)\right]}, \tag{4.22}$$

where $n_\pm$ is the concentration of ionic species. Separating the real and imaginary parts, and taking into account that $\epsilon(\omega) = 1 + q\dot{x}(\omega)/E(\omega)$, we obtain the real part of the dielectric function:

$$\varepsilon'(\omega) = 1 + \frac{n_\pm q^2 \gamma_0}{m\varepsilon_0} \frac{\tau_r b - a}{\omega^2 a^2 + b^2}, \tag{4.23}$$

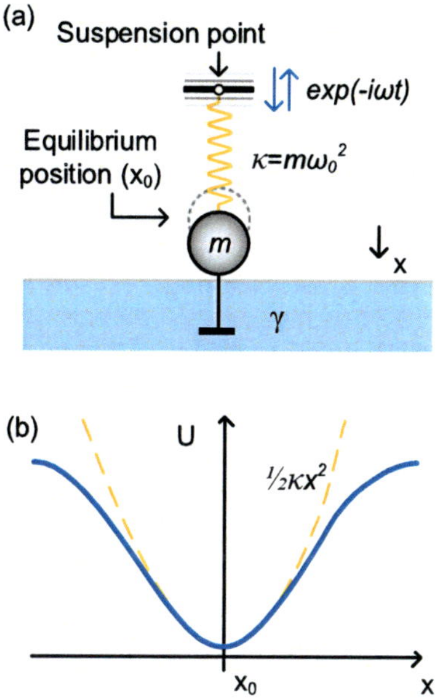

Fig. 4.12 **a** The equivalent scheme for (4.20) explaining the mechanics of the oscillation of ionic species in water in a soft surrounding with friction and moving suspension points. **b** The corresponding potential energy of the ionic species (solid line) and the harmonic parabolic potential approximation (dashed line)

and the real part of dynamical conductivity:

$$\sigma(\omega) = \frac{n_{\pm}q^2\gamma_0}{m}\frac{\omega^2\tau_r a + b}{\omega^2 a^2 + b^2}, \tag{4.24}$$

where $a = \gamma_0(1 + \gamma_0\tau_r)$, and $b(\omega) = \gamma_0\tau_r(\omega_0^2 - \omega^2) + \gamma_0^2 - \omega_0^2$. Note that (4.24) satisfies the sum rule 2.61 (see Sect. 2.8).

Figure 4.13 shows the spectra calculated by (4.23) and (4.24) with different parameters. In the limiting case, $\tau_r = 0$ (free particle) and $\omega_0 = 0$ (oscillator with a zero eigenfrequency), we have a frequency-independent conductivity plateau in all spectral regions. In another limiting case, $\tau_r \to \infty$ (strongly localized charge), the function (4.24) coincides with the Lorentzian oscillator (see curve 7). This case corresponds to a parabolic potential with infinite walls (Fig. 4.12b). In intermediate cases, the solution combines the properties of an overdemped oscillator with static conductivity in the low-frequency region. As the residence time τ_r increases, the low-frequency conductivity decreases, and the peak at ω_0 becomes more pronounced (see Fig. 4.13).

In the case of large damping $\gamma_0 \gg \omega_0$, (4.24) gives the curve with a widened oscillatory peak (see curve 1 in Fig. 4.13). The function has two plateaus, high-frequency and low frequency, which correspond to the quasi-free motion and the hindered motion, respectively. Figure 4.13 shows that at frequencies $\omega \approx \omega_0$, the charge looks more like a harmonic oscillator at the bottom of the parabolic potential, but at low frequencies, $\omega \ll \omega_0$, it overcomes the potential barrier (see Fig. 4.12b) and shows a conductivity plateau:

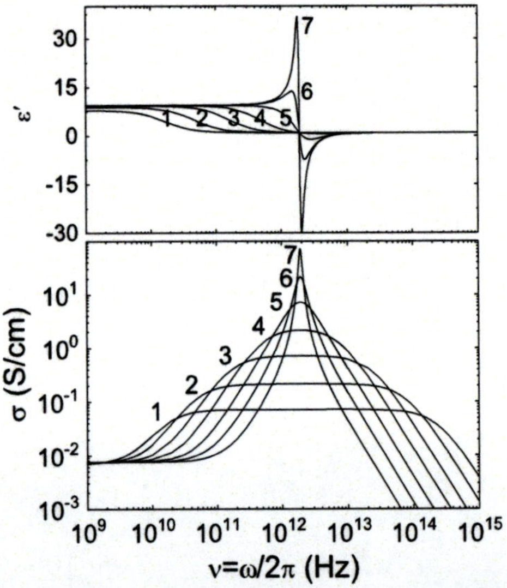

Fig. 4.13 The spectra of the dielectric constant and dynamical conductivity, according to model (4.23) (top panel) and model (4.24) (bottom panel) with the parameters: $n_{\pm} = 1$ mol/l, $\omega_0 = 1.2$ THz, and the friction constant γ_0: (1) 1500, (2) 500, (3) 150, (4) 50, (5) 15, (6) 5 and (7) 1.5 THz

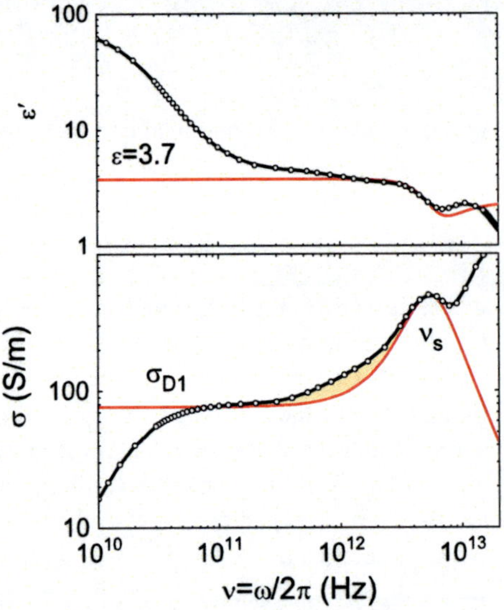

Fig. 4.14 The terahertz part of the dielectric spectrum of water in terms of the real part of the dielectric constant (top panel) and the dynamic conductivity (bottom panel). The red-solid lines are the model according to (4.23) and (4.24) with the parameters: $n_{\pm} = 1$ mol/l, $m = 18 \cdot m_p$, $\gamma_0 = 4$ THz, $\tau_r = 1.6$ ps, and $\omega_0 = 5$ THz

$$\sigma_{D1} = \frac{n_{\pm} q^2 \gamma_0}{m \cdot b}, \tag{4.25}$$

where b is defined above, which corresponds to the steady-state regime of intermolecular charge dynamics without taking into account the interaction between ions (for the case of mutual interaction see Sect. 3.5).

Figure 4.14 shows the experimental spectra of the real part of the dielectric constant (top panel) and dynamic conductivity (bottom panel) of water at room temperature. As one can see, the model (4.24) describes the terahertz part of the spectrum in the 1–10 THz interval, with the parameters given in the figure caption. The mass m of a charged particle is the mass of the ion (H_3O^+ or OH^-), the concentration of charges $n_\pm = 1$ mol/l, and the residence time of the charge in the potential $\tau_r = 1.6$ ps, which coincides with the lifetime t_s of the charge in the oscillatory state (see Table 4.2).

The general terms of the simple model of forced oscillations with friction reproduce the experimentally observed spectrum of water in the terahertz region with the concentration of vibrating species equal to 1 mol/l, as found in the ionic model above. The intense IR spectrum of water is associated with a high concentration of short-lived ionic species, which oscillate in the potential of the surroundings, and also show the translational conductivity as the time of observation increases. The observed deviation of the model curve from the calculated one in the area of the secondary relaxations (see yellow area in Fig. 4.14) is associated with the contribution from the orientation of permanent molecular dipoles, which were not taken into account in the model (4.21). However, the contribution of this part to the static dielectric constant of water $\epsilon(0)$ is only a few percent of the contribution from the main dielectric relaxation, associated with the polarization caused by the relative displacement of short-lived ionic species.

4.5.5 *Autoionization and pH*

The ionic model of water allows us to look at the concept of pH from a new angle. The conductometric method of to determine the dissociation constant K_w of water uses data on the proton DC conductivity of pure water, σ_{dc}, and aqueous electrolyte solutions (see Sects. 1.3.1 and 1.3.3). The dynamics of ions, both of pure water and of dilute electrolytes, are considered independent (no electrostatic interaction between ions is assumed), and the conductivity created by them is supposed to be additive. According to (1.11), the concentration of hydronium ions in pure water is

$$\left[H_3O^+\right] = \sqrt{K_w} = \alpha\,[H_2O] = 10^{pH}, \tag{4.26}$$

where α is a dissociation constant. Despite the simplicity of finding the ion concentration, it has an unusual temperature dependence, $[H_3O^+](T)$, which is unexplained in the Bernal–Fowler model of water.

Figure 4.15 shows the concentration $[H_3O^+]$ of ionic species of pure water as a function of reciprocal temperature (the Arrhenius scale), calculated from K_w(T) shown in Fig. 1.18 by (4.26). Interestingly, the low-temperature activation energy $E_a = k_B T \cdot \ln([H_3O^+]/[H_3O^+]_0)$ of the ionic species (the barrier to the generation of the charge carriers) is only about 0.4 eV, which is an order of magnitude lower than the 5 eV that is expected for the autoprotolysis of water [62], but close to the activation energy of the static DC conductivity of pure water (see Table 2.3). It is

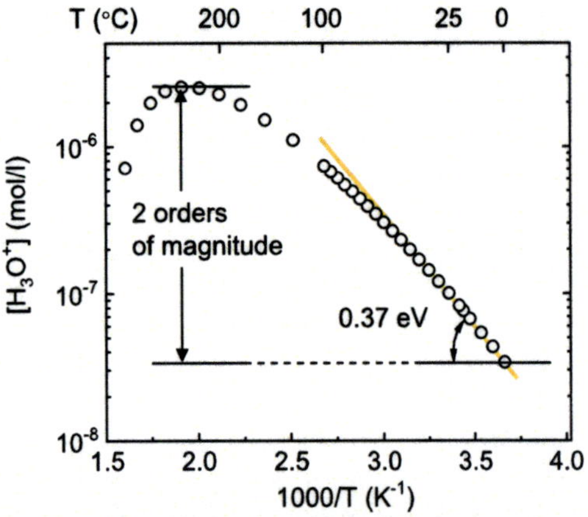

Fig. 4.15 The temperature dependence of the concentration of hydronium ions in liquid water (the Arrhenius scale) in the whole range of the liquid state from the melting point (0 °C) to the critical point (374 °C), calculated according to (4.26). The number near the curve represents the activation energy E_a at low temperatures

also unusual that a relatively small (in the absolute temperature scale) increase of the temperature by about 100 °C causes an increase in the effective concentration of ionic species of almost two orders of the frequency magnitude. Finally, the non-monotonic behavior of the concentration with the maximum at about 250 °C is also rather unexpected.

Figure 4.16 shows the temperature dependencies of the static conductivity of pure water σ_{dc}, the limiting equivalent conductivity Λ_w^0, and the logarithm of K_w, the same as shown in Fig. 1.8. The graph additionally contains the high-frequency (Debye) conductivity, σ_{D1} (see Fig. 3.5), and the corrected value of the limiting equivalent conductivity, $\Lambda_w^{'0}$. The latter was calculated from the limiting equivalent conductivities of electrolyte solutions (Λ_{NaCl}^0, Λ_{HCl}^0, and Λ_{NaOH}^0), accounting for the intrinsic ions of water, which have been added in the right part of each equation in system (1.9) at a concentration of 1 mol/l (see [63] for details). As one can see, $\Lambda_w^{'0}$ is larger than Λ_w^0.

Figure 4.16 also clarifies that the strong temperature dependence of K_w is due to the different temperature dependencies of σ_{dc}(T) and the limiting equivalent conductivity Λ_w^0(T). Although both quantities in fact represent the same mechanism of the transport of intrinsic ions of water, their activation energies 0.4 and 0.1 eV, respectively, are quite different. Note that the latter activation energy is the same for different aqueous electrolytes [64], and also coincides with the activation energy of the high-frequency conductivity of pure water [65]. In other words, the temperature dependency of Λ_w^0 reflects the high-frequency properties of water, where the contribution of the short-lived intrinsic ions was observed [63], rather than the low-frequency static dynamics.

Accounting for the intrinsic ionic species of water, we replace in (1.11) Λ_w^0 with $\Lambda_w^{0'}$, and σ_{dc} by σ_{D1}. Both of them, as seen in Fig. 4.16, have similar temperature dependencies (the same activation energies). At 25 °C, the ratio between $\sigma_{D1} =$ 79 S/m and $\Lambda_w^{0'} = 674\,\mathrm{S \cdot cm^2/mol}$ gives the concentration of ions $n_\pm = 79/0.0674 =$

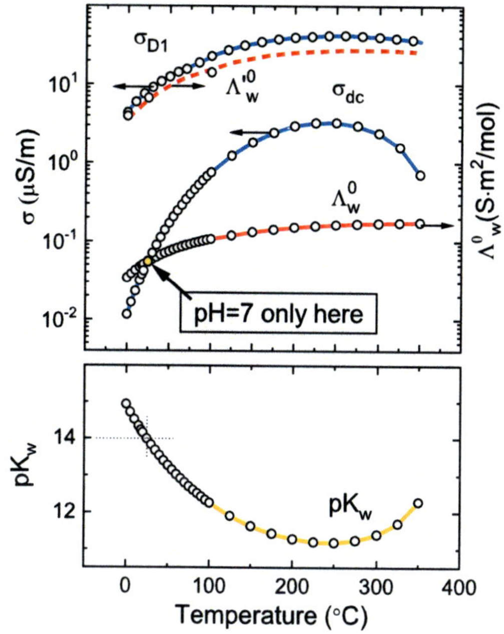

Fig. 4.16 The temperature dependencies from Fig 1.18, and of the high-frequency conductivity of water, σ_{D1}, and the corrected limiting equivalent conductivity, $\Lambda_w^{'0}$. The orange dot shows the point where pH = 7 (see the text for an explanation)

1.1 mol/l, which is equal to the full concentration of intrinsic ions of water (both short- and long-lived) obtained from the analysis of the dielectric spectra (see Sect. 4.2.1).

By analogy with $K_w = [H_3O^+]^2$, where $[H_3O^+]$ is the concentration of long-lived ions only, we can introduce the corrected ionic product $K_w^{'} = n_{\pm}^2$, where $n_{\pm}$ is the full concentration of spontaneously formed ions of water. In the range of 0–100 °C, the latter varies from 1.2 to 1.6. Note that $K_w^{'}$ characterizes the concentration of H_3O^+ and OH^- ions in the dynamic interaction potential, while the standard K_w characterizes the concentration of long-lived ions only. As the temperature increase changes the lifetime distribution of ionic species and effectively extracts them from the potential of mutual interaction, K_w is a measure of the activity of long-lived ions only, while $K_w^{'}$ represents the full concentration of ionic species. In other words, according to the ionic model, the constant K_w (or pH) characterizes the thermal activation of H_3O^+ and OH^- ions from the dynamic potential of their interaction, but not the autoprotolysis of H_2O molecules.

4.6 Concluding Remarks

As follows from Chaps. 2 and 3, the main requirement for a generalized model of the broadband dielectric response of water and ice is a consistent description of the key electrodynamic properties: the high DC conductivity; the high dielectric constant; increased microwave absorption (Debye relaxation); intense IR peaks; the anomalously high mobility of hydronium and hydroxyl ions along with the equal diffusion coefficients of hydrogen and oxygen atoms; and the picosecond proton exchange, which coexist along with the many-hour lifetime of water molecules. The ionic model of water allows one to describe these properties without internal conflicts.

The important innovation of the ionic model is that it accounts for the nuclear quantum effects and ultrashort picosecond-scale chemical reactions, which were previously missing in models of the structure of water (see Chap. 1), by introducing spontaneously formed short-lived ionic pairs with a high concentration $n_{\pm}$ =1 mol/l (2% of the total number of particles, or 1 ion per 50 H_2O molecules). The average distance between ions is $L \approx 1$ nm. This ion–molecular system is highly heterogeneous and dynamic and contains a wide variety of forms of molecular motion and possible ways they manifest themselves in the dielectric response. Due to this spatial-time heterogeneity of water (and ice), their broadband dielectric response is very different in comparison with those for other dielectrics (see Sect. 3.3).

The ionic model creates the prerequisites for resolving the structural and thermodynamic anomalies of water. In particular, the presence of rapidly fluctuating regions of high and low density (see Fig. 4.5) paves the way for a quantitative microscopic description of the anomalously high surface tension, viscosity, and heat capacity of water, which, however, lie beyond the scope of this book. Accounting for the intermolecular interactions of ionic species can help to explain the anomalous melting, boiling, and critical points of water. Moreover, as Röntgen assumed long ago [25], many other anomalous properties of water, such as the non-monotonic behavior of its density, viscosity, compressibility, specific heat capacity, and the speed of sound through it, can be explained if water were a two-component mixture. The ionic model is exactly such a system, where short-lived ionic and molecular species have two sharply defined interaction potentials, and thus water can be considered as a system made of two components, or two types of molecular species. Note, however, that apart from Röntgen's expectations, the species of water are short-lived (picoseconds for water and microseconds for ice) thus the difference between them dissolves on large timescales, as, for example, for the diffusion-averaged D-structure (see Chap. 1). The averaged ion–ion, ion–molecular, and molecular–molecular interactions are presumably also reduced to the mean interactions, which can be construed as a hydrogen bonding. Note also that according to the models discussed in this chapter, the latter concept is inapplicable for the interpretation of the dynamic properties of water on time intervals shorter than 1 μs, and the ionic model should be used.

References

1. E. Arunan, G.R. Desiraju, R.A. Klein, J. Sadlej, S. Scheiner, I. Alkorta, D.C. Clary, R.H. Crabtree, J.J. Dannenberg, P. Hobza, H.G. Kjaergaard, A.C. Legon, B. Mennucci, D.J. Nesbitt, Definition of the hydrogen bond (IUPAC Recommendations 2011). Pure Appl. Chem. **83**, 1637–1641 (2011)
2. E.D. Isaacs, A. Shukla, P.M. Platzman, D.R. Hamann, B. Barbielini, C.A. Tulk, Covalency of the hydrogen bond in Ice: a direct x-ray measurement. Phys. Rev. Lett. **82**, 600–603 (1999)

3. A. Nilsson, H. Ogasawara, M. Cavalleri, D. Nordlund, M. Nyberg, Ph Wernet, L.G.M. Pettersson, The hydrogen bond in ice probed by soft x-ray spectroscopy and density functional theory. J. Chem. Phys. **122**, 154505–9 (2005)
4. P. Atkins, J. de Paula, *Physical Chemistry*, 9th edn. (W. H. Freeman and Company, New York, 2010)
5. W.L. Jorgensen, J.D. Madura, Temperature and size dependence for Monte Carlo simulations of TIP4P water. Mol. Phys. **56**, 1381–1392 (1985)
6. P. Needham, Hydrogen bonding: homing in on a tricky chemical concept. Stud. Hist. Philos. Sci. **44**, 51–65 (2013)
7. B.E. Rocher-Casterline, L.C. Ch'ng, A.K. Mollner, H. Reisler, Communication: determination of the bond dissociation energy (D0) of the water dimer, $(H_2O)_2$, by velocity map imaging. J. Chem. Phys. **134**, 211101–4 (2011)
8. M. Eigen, Proton transfer, acid-base catalysis, and enzymatic hydrolysis. Part I: elementary processes. Angew. Chem. Int. Ed. Engl. **3**, 1–19 (1964)
9. D. Eisenberg, W. Kauzmann, *The Structure and Properties of Water* (Oxford University Press, Oxford, 1969)
10. V.G. Artemov, E. Uykur, S. Roh, A.V. Pronin, H. Ouerdane, M. Dressel, Revealing excess protons in the infrared spectrum of liquid water. Sci. Rep. **10**, 11320–9 (2020)
11. Ya. Frenkel, *Kinetic Theory of Liquids* (Clarendon Press, Oxford, 1946)
12. Y. Huang, X. Zhang, Z. Ma, W. Li, Y. Zhou, J. Zhou, W. Zheng, C.Q. Sun, Size, separation, structural order and mass density of molecules packing in water and ice. Sci. Rep. **3**, 3005–5 (2013)
13. J.D. Bernal, R.H. Fowler, A theory of water and ionic solution, with particular reference to hydrogen and hydroxyl ions. J. Chem. Phys. **1**, 515–548 (1933)
14. T. Hama, H. Ueta, A. Kouchi, N. Watanabe, Quantum tunneling observed without its characteristic large kinetic isotope effects. PNAS **112**, 7438–7443 (2015)
15. L.I. Krishtalik, The mechanism of the proton transfer: an outline. Biochim. Biophys. Acta **1458**, 6–27 (2000)
16. M. Ceriotti, J. Cuny, M. Parrinello, D.E. Manolopoulos, Nuclear quantum effects and hydrogen bond fluctuations in water. PNAS **110**, 15591–15596 (2013)
17. H.J. Bakker, H.-K. Nienhuys, Delocalization of protons in liquid water. Science **297**, 587–590 (2002)
18. H.J. Bakker, J.L. Skinner, Vibrational spectroscopy as a probe of structure and dynamics in liquid water. Chem. Rev. **110**, 1498–1517 (2010)
19. W.B. Carpenter, J.A. Fournier, N.H.C. Lewis, A. Tokmakoff, Picosecond proton transfer kinetics in water revealed with ultrafast IR spectroscopy. J. Phys. Chem. B **122**, 2792–2802 (2018)
20. A. Hassanali, F. Giberti, J. Cuny, T.D. Kühne, M. Parrinello, Proton transfer through the water gossamer. PNAS **110**, 13723–13728 (2013)
21. J.M. Headrick, E.G. Diken, R.S. Walters, N.I. Hammer, R.A. Christie, J. Cui, E.M. Myshakin, M.A. Duncan, M.A. Johnson, K.D. Jordan, Spectral signatures of hydrated proton vibrations in water clusters. Science **308**, 1765–1769 (2005)
22. F. Dahms, B.P. Fingerhut, E.T.J. Nibbering, E. Pines, T. Elsaesser, Large-amplitude transfer motion of hydrated excess protons mapped by ultrafast 2D IR spectroscopy. Science **357**, 491–495 (2017)
23. V.G. Artemov, A. Ryzhov, E. Carlsen, P. Kapralov, H. Ouerdane, Nonrotational mechanism of polarization in alcohols. J. Phys. Chem. B **124**, 11022–11029 (2020)
24. K.S. Singwi, A. Sjölander, Diffusive motions in water and cold neutron scattering. Phys. Rev. **119**, 863 (1960)
25. W.K. Röntgen, Ueber die constitution des flüssigen wassers. Ann. Phys. **45**, 91–97 (1892)
26. A.K. Soper, M.A. Ricci, Structures of high-density and low-density water. Phys. Rev. Lett. **84**, 2881–2884 (2000)
27. K. Amann-Winkel, M.-C. Bellissent-Funel, L.E. Bove, T. Loerting, A. Nilsson, A. Paciaroni, D. Schlesinger, L. Skinner, X-ray and neutron scattering of water. Chem. Rev. **116**, 7570–7589 (2016)

28. J. Frenkel, Continuity of the solid and the liquid states. Nature **136**, 167–168 (1935)
29. J. A. Pople, Molecular association in liquids. A theory of structure of water. Proc. Roy. Soc. **A205**, 163–168 (1951)
30. H.S. Frank, Properties of water and Ice. Proc. R. Soc. Lond. A **247**, 481–492 (1958)
31. T.A. Litovitz, C.M. Davis, Structural and shear relaxation in liquids, in *Physical Acoustics*, ed. by W.P. Mason, vol. 2A (Academic, New York), pp. 281–349 (1965)
32. M.S. Jhon, H. Eyring, Liquid theory and the structure of water. Ann. Rev. Phys. Chem. **27**, 45–57 (1976)
33. D. Chandler, Structure of molecular liquids. Ann. Rev. Phys. Chern. **29**, 441–71 (1978)
34. F.H. Stillinger, T.A. Weber, Hidden structure in liquids. Phys. Rev. A. **25**, 978–989 (1982)
35. J. Higo, M. Sasai, H. Shirai, H. Nakamura, T. Kugimiya, Large vortex-like structure of dipole field in computer models of liquid water and dipole-bridge between biomolecules. PNAS **98**, 5961–5964 (2001)
36. G.G. Malenkov, Structure and dynamics of liquid water. J. Struct. Chem. **47**, 1–31 (2006)
37. M. Eigen, L. De Maeyer, Untersuchungen über die Kinetik der Neutralisation. Z. Elektrochem. **59**, 986–993 (1955)
38. A. Taschin, P. Bartolini, R. Eramo, R. Righini, R. Torre, Evidence of two distinct local structures of water from ambient to supercooled conditions. Nat. Commun. **4**, 2401–8 (2013)
39. C. Huang, K.T. Wikfeldt, T. Tokushima, D. Nordlund, Y. Harada, U. Bergmann, M. Niebuhr, T.M. Weiss, Y. Horikawa, M. Leetmaa, M.P. Ljungberg, O. Takahashi, A. Lenz, L. Ojamäe, A.P. Lyubartsev, S. Shin, L.G.M. Pettersson, A. Nilsson, The inhomogeneous structure of water at ambient conditions. PNAS **106**, 15214–15218 (2009)
40. S. Woutersen, U. Emmerichs, H.J. Bakker, Femtosecond Mid-IR pump-probe spectroscopy of liquid water: evidence for a two-component structure. Science **278**, 658–660 (1997)
41. J.K. Holt, H.G. Park, Y. Wang, M. Stadermann, A.B. Artyukhin, C.P. Grigoropoulos, A. Noy, O. Bakajin, Fast mass transport through Sub-2-Nanometer carbon nanotubes. Science. **312**, 1034–1037 (2006)
42. V.G. Artemov, E. Uykur, P.O. Kapralov, A. Kiselev, K. Stevenson, H. Ouerdane, M. Dressel, Anomalously high proton conduction of interfacial water. J. Phys. Chem. Lett. **11**, 3623–3628 (2020)
43. W. Armstrong, Electrical phenomena. The electrical. J. Phys. Chem. Lett. **10**, 154–155 (1893)
44. E.C. Fuchs, Can a century old experiment reveal hidden properties of water? Water **2**, 381–410 (2010)
45. A. Botti, F. Bruni, S. Imberti, M.A. Ricci, A.K. Soper, Ions in water: the microscopic structure of a concentrated HCl solution. J. Chem. Phys. **121**, 7840–7848 (2004)
46. J. Ditkovich, D. Pines, E. Pines, Controlling reactivity by remote protonation of a basic side group in a bifunctional photoacid. Phys. Chem. Chem. Phys. **18**, 16106–16115 (2016)
47. C. Bai, J. Herzfeld, Special Pairs are decisive in the autoionization and recombination of water. J. Phys. Chem. B. **121**, 4213–4219 (2017)
48. V.G. Artemov, A.A. Volkov, A.V. Pronin, Electrical properties of water: a new insight. Biophysics **59**, 520–523 (2014)
49. L.D. Landau, E.M. Lifshitz, *Physical Kinetics* (Pergamon Press, New York, 1981)
50. N.B. Vargaftik, B.N. Volkov, L.D. Voljak, International tables of the surface tension of water. J. Phys. Chem. Ref. Data **12**, 817–820 (1983)
51. V.E. Vinogradov, Depression of the cavitation centers in water under pulsed tension conditions. Tech. Phys. Lett. **35**, 54–56 (2009)
52. L. Fumagalli, A. Esfandiar, R. Fabregas, S. Hu, P. Ares, A. Janardanan, Q. Yang, B. Radha, T. Taniguchi, K. Watanabe, G. Gomila, K.S. Novoselov, A.K. Geim, Anomalously low dielectric constant of confined water. Science **360**, 1339–1342 (2018)
53. T. Gaudin, H. Ma, A molecular contact theory for simulating polarization: application to dielectric constant prediction. Phys. Chem. Chem. Phys. **21**, 14846–14857 (2019)
54. A. von Hippel, The dielectric relaxation spectra of water, Ice, and aqueous solutions, and their interpret at ion 1, critical survey of the status-quo for water. IEEE Tnans. Electr. Insul. **23**, 801–816 (1988)

55. J.G. Kirkwood, The dielectric polarization of polar liquids. J. Chem. Phys. **7**, 911–920 (1939)
56. L. Onsager, Electric moments of molecules in liquids. J. Am. Chem. Soc. **58**, 1486–1493 (1936)
57. S.W. Peterson, H.A. Levy, A single-crystal neutron diffraction study of heavy ice. Acta Cryst. **10**, 70–76 (1957)
58. E. Greinacher, W. Luettke, R. Mecke, The effect of temperature upon near infrared spectra absorptivities. Z. Elektrochem. **59**, 23–27 (1955)
59. L.D. Landau, E.M. Lifshitz, *Mechanics*, 3rd edn. (Pergamon Press, New York, 1960)
60. P.L. Kapitza, Dynamic stability of the pendulum with vibrating suspension point (in Russian). Sov. Phys. JETP **21**, 588–597 (1951)
61. C.J. Richards, T.J. Smart, P.H. Jones, D. Cubero, A microscopic Kapitza pendulum. Sci. Rep. **8**, 13107–13110 (2018)
62. P.L. Geissler, C. Dellago, D. Chandler, J. Hutter, M. Parrinello, Autoionization in liquid water. Science **291**, 2121–2124 (2001)
63. V.G. Artemov, A.A. Volkov, N.N. Sysoev, A.A. Volkov Jr, On autoionization and pH of liquid water. Doklady Phys. **61**, 1–4 (2016)
64. V.V. Shcherbakov, V.I. Ermakov, Yu.M. Artemkina, Dielectric characteristics of water and electric conductivity of aqueous electrolytes. Rus. J. Electrochem. **53**, 1301–1306 (2017)
65. V.G. Artemov, A.A. Volkov, N.N. Sysoev, A.A. Volkov, Conductivity of aqueous HCl, NaOH and NaCl solutions: is water just a substrate? EPL **109**, 26002–6 (2016)

Chapter 5
Electrodynamics of Aqueous Media

Abstract In the real world water and ice always have boundaries, being confined by the solid or liquid materials, representing the sample in contact with a gas, vacuum, another liquid, or serving as a solvent of electrolyte solution. Interestingly, the properties of such "finite" water, ice, or aqueous media, can differ from those for the theoretical boundless water. In this chapter, we discuss the electrical properties of aqueous electrolytes, the electrodynamics of confined water, the effect of a strong electric field on water viscosity, and the electrification of water droplets, and other systems where the role of boundaries on the properties of water is important, in an external electric field. The questions regarding the role of water in atmospheric phenomena, and the future of the aqueous electrochemical energy systems (fuel cells, supercapacitors, flow batteries, electrolyzers) are discussed.

5.1 The Dielectric Response of Electrolyte Solutions

5.1.1 Why Electrolytes Conduct Electricity

Pure water has been shown to have very low DC conductivity $\sigma_{dc} = 5.5 \times 10^{-8}$ S/cm.[1] However, a small amount of impurities increases σ_{dc} dramatically, whose value can reach up to 1 S/cm, that is, seven orders of magnitude larger than for pure water. What is the mechanism of such a significant change of the electrodynamic properties of water? Many brilliant scientists have been involved in discussions on why and how aqueous electrolytes (dielectrics by nature) conduct electricity. The most crucial debates on this topic took place in the second half of the nineteenth century. Note that at the time, impedance analyzers were relatively new devices and had significantly lower sensitivity, accuracy, and maximal frequency of operation in comparison with what is available now. Figure 5.1 shows the setup used by Kohlrausch for the first systematic studies of aqueous electrolytes. The alternating current from the sine-induction machine allowed him to prevent the deposition of electrolysis products and

[1]It was noted long ago that 1 mm of water at 0 °C has the same resistance as 40 million km of copper of the same cross section [1].

V. Artemov, *The Electrodynamics of Water and Ice*, Springer Series in Chemical Physics 124, https://doi.org/10.1007/978-3-030-72424-5_5

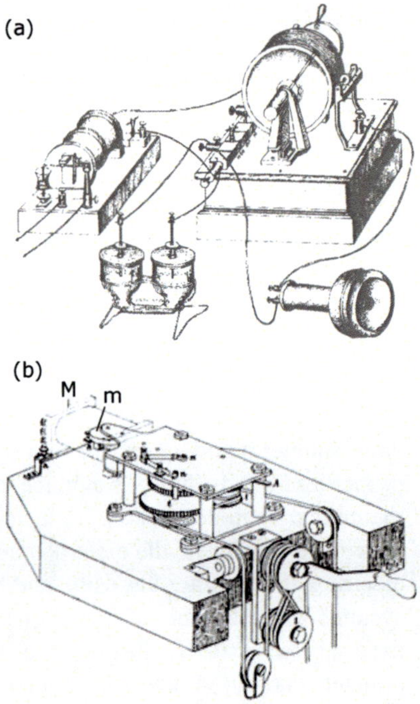

Fig. 5.1 Kohlrausch's setups (1874–1880) for the measurements of the conductivity of electrolytes: **a** An induction apparatus with an iron core, and Neefian hammer (a circuit breaker to produce steady alternating current), and a telephone hook-up [3]; and **b** The sine-induction machine, where *m* is the magnetic plate and *M* is the multiplier [4]

enabled him to obtain precise results, which are still valid today. For example, the modern textbook value of DC conductivity of the pure water is still that obtained by Kohlrausch. However, due to technical limitations, the first electrolytic measurements were limited to a frequency typically around 1 kHz. Thus, one should understand that the basic mechanism of the conduction of electrolytes was developed on the basis of low-frequency measurements. In other words, the structure of electrolytes was assumed on the basis of the frequency region where water shows no dispersion, demonstrating a stationary DC plateau (see Fig. 2.2).

The frequency-independent static electrical conductivity of electrolytes led early investigators of aqueous solutions to the conclusion that there are charge carriers, which transfer electricity, being drugged by the external electric field, and, thus, producing current that depends on the concentration of the solute.[2] The role of water in the electric-current conduction mechanism was assumed to be secondary. In fact, water was considered as a substrate, which separates the solute, and provides the medium of the effective translation of the charged particles. Remarkably, the understanding that electrolytes and metals have different types of the conduction mechanisms did not come immediately. In particular, Nernst [2], discussing the conduction of metals by analogy with electrolytes, assumed that metals also contain the products of dissociation, but the conducting particles are mass-less, and presented in a very high concentration, thus providing a significantly larger current. In other

[2]Note that the difference between atoms and molecules was not yet fully understood at that time.

words, metals were initially considered as solvents with a high dissociation rate by analogy with aqueous electrolytes. The modern understanding of metals came much later, with the advent of quantum theory.

In the historical context, it is also interesting to note other ideas [2], which significantly determined our understanding of electrolytes, but have also misled researchers for decades. It was assumed that if the electrolyte consisted only of charged particles, then their separation under the influence of an electric field would lead to their scattering in space. Since this was not observed, only complex substances (not elements) were endowed with the ability to conduct electricity. It was found that ions in the solutions and in the melts of a homogeneous substance have the same electric charge, thus, both contain similar species of the same nature. Finally, the conductivity of salt melts was found proportional to the degree of dissociation, but inversely proportional to the friction, and the separation of these two factors by the measurements of the conductivity only was found to be impossible.

Kohlraush was probably one of the most productive experimentalists of that period. In 1874, he demonstrated [4] that each electrolyte has a definite solute-specific electrical resistance. By measuring the conductivity of different aqueous electrolytes on dilution (see Fig. 5.2), he derived a simple dependence of the equivalent conductivity Λ on the concentration c:

$$\Lambda = \frac{\sigma_{dc}}{c} = \Lambda_0 - A\sqrt{c}, \tag{5.1}$$

where Λ_0 is the equivalent conductivity at infinite dilution, and A is the electrolyte-specific constant. It has been shown that (5.1) perfectly describes the static electrical conductivity σ_{dc} of all strong electrolytes up to the concentration of about 1 mol/l. The full concentration of molecular species in water is 55.5 mol/l. Kohlrausch then derived the law of the independent migration of ions:

$$\Lambda_0 = n_+\lambda_+ + n_-\lambda_-, \tag{5.2}$$

where λ_+ and λ_- are the limiting molar conductivities of cations and anions, respectively, and n_+ and n_- are the stoichiometric numbers of positive and negative ions formed during the dissociation of the electrolyte. Analyzing the static conductivity of electrolytes at relatively large dilutions, he concluded that each type of migrating ion has a specific limiting molar conductivity no matter what combination of ions are in solution, and therefore that a solution's electrical resistance is due only to the migrating ions of a given substance. Thus, the idea of the independent dynamics of long-lived ionic species appeared. Later, analyzing the law for electrolytes of higher concentrations, Debye and Hückel found that there were significant deviations between theory and experiment at the high concentration limit, and suggested electrophoretic and relaxatory approximations, accounting for the interaction between ions and between ions and the solvent (water), respectively. Although these corrections minimized the difference between theory and experiment, the exact mechanism of charge transfer in concentrated electrolytes still lacked clarity. The corresponding

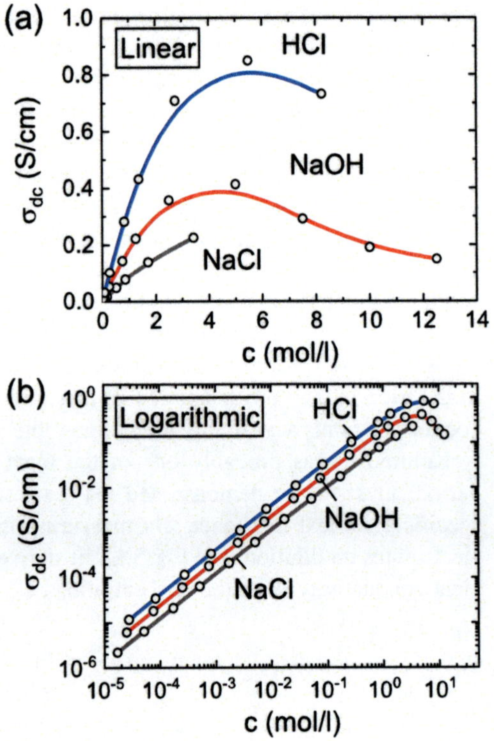

Fig. 5.2 The static electrical conductivity of an aqueous solutions of NaCl (salt), NaOH (base), and HCl (acid) as a function of the concentration, c, in: **a** linear, and **b** double-logarithmic scales. The linear behavior at low concentrations is in accordance with Kohlrausch's law (5.3)

Debye–Hückel–Onsager equation:

$$\Lambda = \Lambda_0 - (A + B\Lambda_0)\sqrt{c}, \tag{5.3}$$

where A and B are constants that depend only on known quantities such as temperature, the charges on the ions and the dielectric constant and viscosity of the solvent, allowing one to describe the experimental data with higher accuracy than the original Kohlrausch equation (5.1). However, (5.3) also satisfies the experimental data at relatively low concentrations only, and an interpretation on the microscopic level was also needed.

In his dissertation, Arrhenius [5] suggested considering the disassociation of solid crystalline salts into paired charged particles when they dissolved in water,[3] Arrhenius's explanation was naive but simple and clear enough to win him a Nobel prize in Chemistry in 1903. In this way, the notion of "ions," given by Faraday to

[3]There is a legend that Ostwald commented on the Arrhenius dissertation in the summer of 1884 in the following way: "I spent a feverish night with bad dreams. At the same time I had a severe toothache, a newborn daughter, and Arrhenius's article 'Electrolyte Conductivity Studies'. What was written in the work was so different from the familiar that at first I was inclined to take everything as a whole for nonsense."

the charge carriers in electrolytes many years earlier, firmly entered the history of the physical chemistry of solutions. Apart from Faraday's assumption that ions are produced by an external electric field, Arrhenius proposed that, even in the absence of an electric field, solutions contain ions. He thus introduced the idea that chemical reactions in solution were reactions between ions only. No water molecules were involved in his consideration of the electric current at all. According to his theory, aqueous electrolytes conduct due to the presence of long-lived free ions. The idea that more free ions provide more current was in line with the ideas of Kohlrausch.

Interestingly, neither Kohlraush nor Arrhenius ever saw the Debye relaxation of water, which appears at frequencies higher than those available at that time and was discovered years later. As the optical transparency window (see Fig. 2.2) is almost on the same level as DC conductivity, they probably even did not expect that a huge absorption band exists in the gigahertz-to-IR region. Nor did they know that the microwave part of the spectrum is almost unaffected by the electrolyte concentration (see Sect. 5.1.2), and thus were unable to include high-frequency mechanisms in their considerations. Maybe if they had, the theory of electrolytes would be more general and completely different.

Mendeleev did not agree with the postulates by Arrhenius [6, 7]. His opinion was based on the experience of chemists, who dealt with chemical kinetics, and the idea of the physical concept of the electrolytes, where ionic species exist as spherical particles and do not interact with the solvent (water) was controversial. He noted [8] that rather than accepting that for the salt MX the dissociation of its particles into ionic species M+X, we should first, following our knowledge on the nature of electrolytes, look for aqueous electrolytes of the salt MX in their interaction with H_2O, which gives MOH + HX. In other words, Mendeleev's idea was to account for local chemical reactions and the chemical exchange between the species of solute and solvent. His ideas and arguments were actively supported by Armstrong, Crompton, Pickering, Traube, Wiedemann, and others. Nevertheless, Arrhenius [9], enlisting the support of Ostwald and Van 't Hoff, defended himself by the argument that the isotonic coefficient i[4] can be explained by his theory of electrolytic dissociation. Providing his arguments, he only rejected that part of Mendeleev's idea related to clathrates but was not able to appreciate the pivotal idea of the ion–molecular exchange interaction proposed by Mendeleev, which would significantly move forward the understanding of the dynamic structure of aqueous electrolytes.

There several points that show the imperfection of the theory of Arrhenius (and all subsequent theories on this basis). First, the theory does not explain the mechanism of dissolution. It operates with the already dissolved ionic species without explanation of where water gets the huge amount of energy (hundreds of kJ/mol) required to separate the ions of the initially neutral solvent. The second problem is that the theory works for very diluted electrolytes (usually below 0.001 mol/l) but fails to reproduce results for higher concentrations (even accounting for the Debye–Hückel amendments [10]). The third problem is that the model of Arrhenius does not account

[4]The isotonic coefficient (or van 't Hoff's coefficient) is the proportionality coefficient between the total concentration of a solute and the concentration of solute particles.

for the mechanisms of solvation,[5] which, as shown by IR spectroscopy [11], is very important for the dynamic structure of electrolytes. The fourth problem is that the theory gives large mobilities for large ions [12]. For instance, the ions of Li^+, Na^+, K^+, Rb^+, Cs^+, etc., are of the same charge, and ordered by the increasing crystallographic radius, and, consequently, their mobilities should gradually decrease from left to right. However, their mobilities in aqueous solutions appear to show the opposite behavior, demonstrating an increase of mobility as the size increases. Although this effect is usually explained by smaller ions getting larger solvation shells, such an assumption contradicts the general physical principles of electrostatic interactions. Finally, the theory is only valid for the static DC current and says nothing about the frequency dependence of the conductivity. In particular, it does not explain the similar microwave absorption of pure water and strong electrolytes. All these points can be accounted for, but the initially simple model was getting so complicated that its practical application is complicated. As a result, in practice, the transfer phenomena in aqueous solutions (including electrical properties) are described by the empirical coefficients [12], while the real atomic-molecular structure of aqueous electrolytes is missing.

Summarizing, at the beginning of the twentieth century, the mechanism of the conductivity of electrolytes was better understood than the mechanism of electrical conduction in metals. The simplicity of Arrhenius' ideas found many supporters, and his ideas were transformed into the modern theory of solvation, which, however, still has some open questions, as described above. Nowadays everything is upside down, while the theory of metals has been significantly elaborated for decades [13], the structure and dynamics of aqueous electrolytes have not changed much in comparison with the knowledge that we had at the beginning of the twentieth century. Interestingly, earlier views on the structure of aqueous electrolytes are still very popular in textbooks, and the basic understanding of the kind of phenomenology discussed above is still used for the description of many electrochemical processes. However, recent progress in quantum chemistry, computational methods, and especially femtochemistry require a significant reconsideration of the quasi-static approach developed more than hundred years ago.

5.1.2 The Frequency-Dependent Conductivity of Aqueous Electrolytes

Hasted et al. conducted [14] the first systematic study of the frequency-dependent dielectric function of different aqueous electrolytes. The results are assembled in [15]. In the past decade, the accuracy of the data and the frequency coverage have been significantly increased and this has allowed us to systematize the electrodynamic data of the basic aqueous electrolytes from the static parameters, such as the dielectric constant and DC current, up to the terahertz region. Figure 5.3 shows the frequency

[5] Solvation describes the interaction of the solvent with the dissolved molecules.

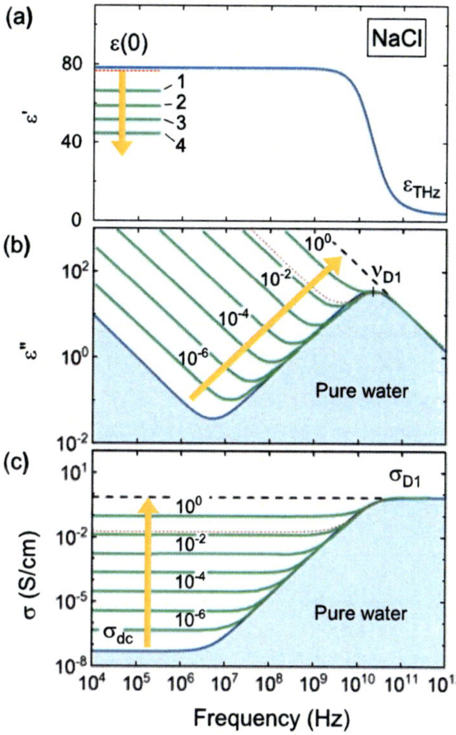

Fig. 5.3 The dielectric spectra of water (blue) and an aqueous solutions of NaCl (green) in terms of: **a** the real part of the dielectric constant (ϵ'); **b** the imaginary part of the dielectric constant (ϵ''); and **c** the dynamic conductivity. The numbers near curves show the concentrations in mol/l. The red-dashed lines are the spectrum of the human blood plasma, which has a salinity of 0.154 mol/l. The yellow arrows show the direction in which the spectra changes as the concentration increases

dependence of the dielectric parameters: conductivity σ, the real ϵ', and imaginary ϵ'', parts of the dielectric constant for NaCl solutions of different concentrations from 10^{-6} mol/l to 1 mol/l (see green lines).[6] The spectra for human blood plasma (dashed red line), and the pure water (blue lines) are given for comparison in order to correlate the data with those for known objects. The data for water serves as a reference spectra. As one can see, the electrolyte added to water changes its dielectric spectrum. DC conductivity is very sensitive to the presence of electrolytes. It increases with the increasing concentration, while the high-frequency part (the Debye relaxation region) stays roughly the same as that for pure water. In the given interval of the concentrations, there are no significant shifts in the relaxation frequency ν_{D1}, no broadening of the relaxation band, and no visible changes in the level of absorption of microwaves and in the terahertz region. At higher concentrations ($c > 0.1$ mol/l), however, the dielectric relaxation transforms, causing a decrease of the dielectric constant $\epsilon(0)$, followed by a decrease of the high-frequency conductivity plateau σ_{D1}.

[6]Interestingly the human taste threshold for salt is about 2×10^{-2} mol/l [16], which is many orders of magnitude higher than the sensitivity of current electronic devices.

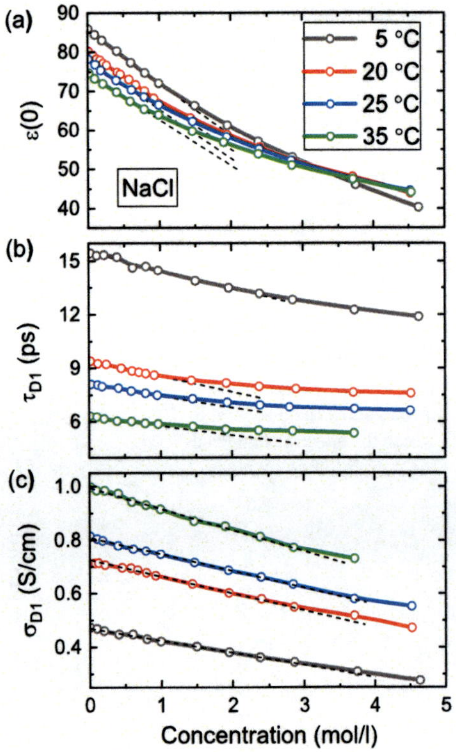

Fig. 5.4 The concentration dependencies of: **a** the static dielectric constant $\epsilon(0)$, **b** the relaxation time τ_{D1}, and **c** the high-frequency conductivity $\sigma_{D1} = \Delta\epsilon\epsilon_0/\tau_{D1}$ of an aqueous solution of NaCl at different temperatures (5, 20, 25, and 35 °C). Dashed lines show the linear dependencies at low concentrations. The experimental data are from [14] to [18]

Figure 5.4a shows the dependence of the dielectric constant $\epsilon(0)$ of a NaCl solution on the concentration at different temperatures. A significant decrease of the static dielectric constant with an increase of the salt concentration has been called dielectric decrement (dielectric depression) [14]. This effect is quite unusual, as according to Arrhenius' theory, the number of charges should increase with the concentration, and, as a result, the additional polarizability should contribute to the dielectric constant, while the experimental dielectric constant of aqueous electrolytes shows the opposite behavior. Intuitively, the dielectric decrement stems from the fact that the local electric field generated by the solute inhibits the external applied field, thus reducing the dielectric constant (see Sect. 2.4.2). In dilute solutions (typically with the concentration less than 1 mol/l) the dielectric decrement is linear (see Fig. 5.4a):

$$\epsilon(0) = \epsilon_w(0) - \alpha c, \tag{5.4}$$

where $\epsilon_w(0)$ is the dielectric constant of pure water, and α is a phenomenological electrolyte-specific coefficient. At high concentrations a significant deviation from linear behavior (5.4) is observed as if the "efficiency" with which the solute counteracts the external field would decrease with an increase in concentration.

It has been found [17] that the dielectric decrement becomes saturated (takes the value of about half that for pure water) at the same concentration point where the DC conductivity reaches its maximum (see Fig. 5.2a). Interestingly, the level of maximal static electric conductivity σ_{dc}^{e} of all strong electrolytes is always higher than the σ_{dc} of pure water, and below, but close to, the high-frequency conductivity plateau $\sigma_{D1} \approx 1$ S/cm of pure water (see the dashed black line in Fig. 5.3c). This looks as if the high-frequency dynamic structure of water would establish the limit for the maximal conductivity of electrolytes.[7] As in the ionic model of water (see Chap. 4), we associated the plateau, σ_{D1}, with the conductivity of spontaneously formed short-lived ionic species (H_3O^+ and OH^-) of water, the fact that the maximal conductivity of the electrolyte is observed when the solute concentration is close to the concentration of intrinsic ions, indicates that the species of solute and solvent interact with each other, as was highlighted by Mendeleev [8]. Further details are explained in Sect. 5.1.3.

Note that the dielectric decrement was not considered in Arrhenius' model, as it was not observed until later. For the past decades, several ideas on how to incorporate this effect into the concept of electrolytic dissociation have been suggested by Haggis, Hasted, and others. Haggis et al. [19] modeled the observed linear dielectric decrement by assuming that the water molecules, adjusting to the electric field of the ion of the electrolyte, create a small spherical region (hydration shell) with a smaller dielectric constant than the rest of the water. The effective medium, in which some of the H_2O molecules are trapped in the hydration shell and, thus, cannot effectively respond to the external electric field, was shown qualitatively to explain the static dielectric constant. However, such theoretical deductions suffer from the introduction of arbitrary assumptions such as the saturation level of the hydration shell, with no chance to give a quantitative description, although the right order of magnitude of the dielectric constant of solutions can be obtained.

In fact, the electric field of a charge with the size of a monovalent ion, placed in a polarizable environment, is a way that the first shell is far from saturated. Glueckauf suggested [20] improving the Haggis's model by introducing variation in the local dielectric constant near the ion. As this approach gave better but still not perfect results, a further improvement was suggested by Liszi et al. [21] to account for the exclusion volume by ions and the corresponding finite-size effects. Although the correct values of the dielectric constants were finally obtained, the model still contains poorly defined coefficients, whose microscopic physical meaning lack clarity. The main problem is that the theory of Arrhenius works well for the static dielectric

[7] Note that apart the very specific cases of superacids and superbases, the pH values of aqueous electrolytes lie in the interval between pH = 0 and pH = 14, with the middle point of pH = 7 for the pure water. By definition, the concentration of $[H_3O^+] = 10^{-pH}$, the maximal difference in the concentration of hydronium or hydroxide ions with those in neutral water is seven orders of magnitude (7 is exactly in the middle of the 0 and 14). The difference between the low-frequency DC plateau of pure water and its high-frequency plateau is also seven orders of magnitude. Thus, the difference in between σ_{dc} and σ_{D1} of pure water (see Fig. 5.2c) sets the dynamic range of the pH of all aqueous solutions of electrolytes.

parameters only, ignoring the frequency dependencies of the dielectric response, which defines the static dielectric constant according to (2.14) and missing the natural dynamism of the water-electrolyte molecular system.

Figure 5.4b shows the dielectric relaxation time $\tau_{D1} = 1/(2\pi\nu_{D1})$ of a NaCl solution as a function of concentration. The small effect of the electrolyte on the ν_{D1} of water is observed up to the saturation point (about 6.1 mol/l). However, one can unambiguously identify the blue shift of the relaxation maximum. In Chap. 4, we associated the relaxation time with the lifetime of molecules of H_2O. In the context of this shift, we can assume that the averaged lifetime of H_2O molecules in water decreases as the electrolyte concentration increases, which, according to (4.1), leads to an increase of the effective concentration $n_{\pm}$ of intrinsic ions. Thus, according to the ionic model, the lifetime distribution of the intrinsic ionic species of water is affected by the species of electrolyte, once again hinting at their interaction.

Further, according to the Debye formula (2.38), high-frequency terahertz conductivity is defined as $\sigma_{D1} = \Delta\epsilon\epsilon_0/\tau_{D1}$, where $\Delta\epsilon = \epsilon(0) - \epsilon_{THz}$. The latter is about 3 (see Fig. 2.4). Figure 5.4c shows the concentration dependence of the plateau σ_{D1}, calculated by the Debye formula. As one can see, σ_{D1} decreases as the concentration increases, while the DC conductivity shows the opposite behavior (compare with Fig. 5.2a). It is interesting to analyze this fact in the context of the conductivity sum rule discussed in Sect. 2.8. The rule states that the total integral of the dynamic conductivity function (the area under the curve) is conserved and depends on the number of charge carriers per unit volume. The decrease of the high-frequency conductivity plateau and the increase of the low-frequency DC conductivity plateau can be understood as the transfer of the dielectric contribution from high frequencies to low frequencies. In other words, the particles of the electrolyte affect the dynamic structure of water in such a way that the intrinsic ionic species, which contributed only at high frequencies due to their mutual interactions, extend their contribution into the low frequencies, while the total contribution to the integral (2.63) is conserved. This effect of communicating high- and low-frequency plateaus can explain the observed synchronous decrease of σ_{D1} and increase of σ_{dc} with concentration.

Note that the increase of the conductivity of electrolytes at high frequencies of the external field (see Fig. 5.3) was predicted by Falkenhagen [22] (the Debye–Falkenhagen effect). He associated the phenomenon with the mutual screening of ionic species, which have low conductivity at low frequencies as they are locked by the potential of electrostatic interaction. According to Falkenhagen, as the frequency of the external field increases, the ionic species interact with the field more effectively, as they can move inside the potential (the rattle effect), and, thus, show higher conductivity. The conductivity of the electrolyte increases starting from frequencies of about 10 MHz (see Fig. 5.3c), however, Falkenhagen's analysis did not show that the slope repeats exactly those observed for pure water, which would not show the increase of the conductivity according to his assumptions. In other words, modern data show that the spectrum looks more like the intrinsic species of water (not an electrolyte) interacting with the external field. Further, the frequency dispersion of the conductivity is better considered as a result of the interaction of the solvent and solute, not as an intrinsic property of the solute itself.

More interesting facts related to the frequency-dependent electrodynamic parameters of electrolytes were found by Scherbakov et al. [23]. It has been shown that the low- (σ_{dc}) and high-frequency (σ_{D1}) conductivity of aqueous electrolytes are correlated and that the activation energy of DC conductivity coincides with the activation energy of σ_{D1} within experimental error. The activation energies of the high-frequency plateaus were found to have the same value of about 0.15 eV for a wide range of aqueous electrolytes. In other words, the thermodynamic properties of the high- and low-frequency conductivity plateaus assume that they are connected and both represent the same mechanism, but this mechanism is averaged over the picosecond period for the high-frequency plateau and the microsecond period for the low-frequency plateau. This fact confirms the results of the analysis within the sum rule (see above).

Finally, the analysis of the frequency-dependent dielectric function and dynamic conductivity of aqueous electrolytes in Arrhenius' approach leads to the paradox of negative hydration [24, 25]. The problem appears when we calculate the mobility of ions from their conductivities according to the Stokes–Einstein equation [12]. The effective radius of an imaginary hydrated ion (for simplicity, let us consider only those from the first column of the periodic table, namely, Li^+, Na^+, K^+, Rb^+, Cs^+, and Fr^+, and, thus, having the same electric charge) decreases as the ion size increases. As follows from the electrochemical mobility measurements, the effective radius of the hydrated ion changes several times from ion to ion [12]. This result is strange, as the real (crystallographic) radius of ions does not change much, and similar (by size and charge) ions should create similar electric fields, which should attract similar numbers of H_2O molecules. Moreover, the effective hydration shells of ions of the large masses, such as Cs^+ and Fr^+, are smaller than their crystallographic radius. This is nonsense. This effect also finds some hand-waving interpretations [25], the paradox obtained from the theory of electrolytic dissociation is evident and suggests that basic assumptions of the initial model are inaccurate.

Thus, experimental facts that were unavailable when the theory of electrolytic dissociation emerged are beyond Arrhenius' generalized concept. These results belong to high-frequency data and assume a dynamic interaction between solvent and solute, which is missing in the static theories. The new phenomenological approach considered below extends the validity of the ionic model of water (see Chap. 4) from pure water to the aqueous solutions, providing a new vision of the dynamic structure of aqueous electrolytes, accounting for picosecond local chemical reactions. Although this concept needs to be verified and further developed, the general results allow us to conclude that the basic assumptions are promising and can be used for the elimination of the anomalies associated with water and aqueous electrolytes.

5.1.3 The Mechanism of Dynamic Conductivity in Electrolytes

In Chap. 4, we modeled the spectrum of the dynamic conductivity of water using the notion of short-lived excess protons, which spontaneously appear and disappear on the picosecond timescale, while some of them can live longer, contributing to the static DC conductivity. Although most ions are short-lived, their concentration up to 1 mol/l (see Fig. 4.5 for the approximate proportions between ions and water) is high enough to induce an electrostatic field, which significantly determines the properties of water at high frequencies (above 1 MHz) and, as it shown in Sect. 5.2, at nanoscales. Such a consideration makes water a strong intrinsic electrolyte with electrophoretic effects (among others) of ion–ion and ion–molecule interactions, which are usually missing in the models of aqueous electrolytes. Here we avoid the analysis of the standard models, which can be found in a variety of textbooks [12, 26–30] and consider the model which accounts for the full dielectric spectrum of aqueous electrolytes from the DC plateau to the far-IR region.

The implementation of intrinsic short-lived ionic species to the water model assumes interactions between them and the particles of solute both electrostatically and via exchange reactions. For simplicity, we consider the aqueous solutions of HCl (acid), NaOH (base), and NaCl (salt) in order to embrace the full spectrum of possible interactions of the solute with water particles. However, the model can be easily extended to a wide range of aqueous electrolytes. The initial particles of the solute are short-lived HCl, NaOH, and NaCl. In the ionic model, these molecular species can interact with molecules of H_2O, forming ions, such as Na^+ and Cl^- (note that Arrhenius' model postulated the complete dissociation of electrolyte molecules), but in addition can also interact with the intrinsic ions of water, H_3O^+ and OH^-, forming molecules of HCl and NaOH. A further development of the model assumes the identification of corresponding reaction rates between all these species.

For clarity, we distinguish the intrinsic water ions, H_3O^+ and OH^-, and the H_3O^* and OH^* ions which have just been born due to the electrolyte-water reaction [17]. The ions of these two types differ by their lifetimes and by their contribution to the dielectric spectrum. While the first pair respond to the high-frequency conductivity plateau σ_{D1} (see Fig. 5.3c) the second pair are born initially free of the electrophoretic effect. In fact, the HCl and NaOH molecules, being neutral, do not feel the interionic potential. Since they generate the H_3O^* and OH^* ions in a random way, the averaged potential for these ions is flat in comparison with that for the intrinsic ions. Their conductivity spectrum can be considered as a frequency-independent horizontal line (see Fig. 4.8), whose altitude (the level of conductivity) depends on their concentration and reaches the maximal values of the dashed black line in Fig. 5.3c. Thus, the conductivity produced by the ions born due to the chemical reaction with the solute (H_3O^* and OH^*) varies in the limits from the σ_{dc} of the pure water to the σ_{D1}. In other words, we consider the activation of ionic species from electrophoretic suppression by the electrolyte and their manifestation in DC conductivity.

Note that the ions of the electrolyte Na^+ and Cl^- drop out of the time scale of our consideration. At the scale of chemical activity, the Na^+ and Cl^- ions are to the left hand of the H^+ and OH^- ions and therefore, like protons, form a chemical bond on femtosecond timescales.

We consider the aqueous NaOH solution as an example to further extend the model of the HCl and NaCl solutions. We suppose that

1. The NaOH in the solution is decomposed completely, or, in other words that its initial concentration $c_0^e = [NaOH]_0$ and the effective concentration $c^e = [NaOH]$ are equal ($[NaOH]_0 \approx [NaOH]$);
2. The solution is electrically neutral ($[H_3O^+] = [OH^-] + [OH^*]$);
3. The total number of protons (hydrogen nuclei) conserves ([H] = const), and the concentrations of ionic species are small in comparison with the initial concentration c_0 of H_2O molecules in pure water ($c_0 - c_0^e \approx [H_2O] + c^e$).

There are two interaction reactions: between the ions and the water molecules, and between the solute and the ions of water, which can be characterized by the following laws of mass actions:

$$K_1 = \frac{[H_3O^+]\left([OH^-] + [OH^*]\right)}{[H_2O]^2}, \tag{5.5}$$

and

$$K_2 = \frac{[OH^-]}{[OH^*]} \cdot c^e, \tag{5.6}$$

wheres K_1 and K_2 are equilibrium constants of reactions, and c^e is the effective concentration of NaOH (e = electrolyte) in the equilibrium.

Equations (5.5) and (5.6) give the concentration of the DC-active ions according to the Nernst–Einstein relation:

$$\sigma_{dc}^e = \frac{q^2}{k_B T}[OH^*]D_{OH^*}, \tag{5.7}$$

where q is the elementary charge, k_B is the Boltzmann constant, T is the temperature, and D_{OH^*} is the diffusion coefficient of OH^* ions.

Substituting (5.5) and (5.6) to (5.7) we get [17]:

$$\sigma_{dc}^e = \frac{q^2}{k_B T} D \frac{\sqrt{K_1}}{c_0} \frac{\left(c_0 - 2c_0^e\right)^2}{K_2 + K'_2 c_0^e} c_0^e, \tag{5.8}$$

where c_0^e is the initial concentrations of NaOH. Although (5.8) was derived for NaOH, it is also valid for HCl. As this electrolyte acts on water mirror-symmetrically, we should simply replace OH^- ions with H_3O^+. In the limiting case when $c_0^e \to 0$, (5.8) transforms to the expression for the equivalent conductivity at infinite dilution Λ_x^0, similar to the empirical (5.1) derived by Kohlrauch:

$$\Lambda_0^e = \sigma_{dc}^e / c_0^e = \frac{q^2/k_B T \cdot D\sqrt{K_1 c_0}}{K_2}. \quad (5.9)$$

To make the model valid for NaCl (salt), we need to add one more equation to (5.4) and (5.5), which accounts for the local chemical reactions that is absent in HCl (acid) and NaOH (base) systems:

$$K_4 = \frac{[HCl][NaOH]}{[NaCl]}, \quad (5.10)$$

where K_4 is a constant that characterizes the neutralization of the ionic species of the solute by the ionic species of water, which produces neutral molecules of HCl and NaOH (or, in the more general case, the species of MOH and HX, where the MX is the initial chemical formula of the solute). In other words, the ionic model differs from the Arrhenius model by accounting for the chemical reactions between solute and solvent (water), which were previously missing. For example, if the initial solute was made of the molecular species MX, Arrhenius' model assumes the following reaction of dissociation:

$$MX \rightarrow M^+ + X^-, \quad (5.11)$$

where M^+ and X^- are the ionic products of dissociation, while the ionic model supplements this equation with the following two equations of protonation:

$$M^+ + OH^- \leftrightarrow MOH, \quad (5.12)$$

and

$$X^- + H_3O^+ \leftrightarrow HX + H_2O, \quad (5.13)$$

which were not included in the previous considerations, as water was supposed to be a neutral molecular system with negligibly small concentration of intrinsic ions (see Chap. 1 for critical analysis).

To simplify the further analysis we modify the reaction in points (1) to (3) above as follows:

1. $[NaCl]_0$ = [NaCl] + [HCl] + [NaOH], where we assumed that the rate of reactions (5.12) is roughly the same as that for reaction (5.13), or in other words that [HX] ≈ [MOH];
2. $[H_3O^+] + [H_3O^*] = [OH^-] + [OH^*]$;
3. $c_0 \cdot [NaCl] = [H_2O] + [HCl] + [NaOH]$.

As the effective volume, occupied by the products of reactions, could differ from the initial volume of reactants, we introduce the corresponding coefficient θ and write [NaCl] = ([HCl] + [NaOH])θ or, in the different form, accounting for (5.9):

$$\sigma_{dc}^{NaCl}\frac{1}{\Lambda_0^{NaCl}} = \sigma_{dc}^{NaCl}\left(\frac{1}{\Lambda_0^{HCl}} + \frac{1}{\Lambda_0^{NaOH}}\right)\cdot\theta, \tag{5.14}$$

where θ is an electrolyte-specific parameter, which for NaCl equals 1.2, as follows from the substitution of Λ_0^e from the textbook [12]. Note that this coefficient is close to the change of the density of saturated NaCl in comparison with that for pure water ($\rho_{NaCl}/\rho_w \approx 1.19$). Furthermore, $K_4 = [4\theta(1+\theta)]^{-1}\cdot c_0^{NaCl}$ and the proportion for the concentrations for HCl, NaOH and NaCl is 1:1:2θ.

Equation (5.8) can be adapted for the NaCl solution by substituting c_0^e for βc_0^{NaCl}, where $\beta = \theta/(1+\theta)$. Finally, the general formula for the contribution of the electrolyte to the DC conductivity of water is

$$\sigma_{dc}^{e} = \frac{q^2}{k_B T}\frac{\Lambda_0^e K_2}{c_0^2}\frac{\left(c_0 - \alpha\gamma c_0^e\right)^2}{K_2 + K'_2\beta c_0^e}\beta c_0^e, \tag{5.15}$$

where α equals 2 for acids and bases, and <2 for salts, and reflects the difference in points (1) to (3) between these types of electrolytes, and γ is a correction factor, which reflects the change of the volume of the solution in comparison with the initial water volume (the expansion/reduction factor).

The static dielectric constant of water is $\epsilon(0) = \Delta\epsilon_w - \epsilon_{THz}$, where $\epsilon_{THz} \approx 3$, and $\Delta\epsilon_w$ is the dielectric contribution of the main relaxation, which according to (4.3) and (4.4) is proportional to $(n_\pm)^{-1/3}$, where $n_\pm$ is the full concentration of ionic species which contribute to conductivity at high frequencies. In aqueous solutions, the concentration of ionic species, according to (5.5), is $n_\pm = (2\sqrt{K_1}c_0 - c^e)n_\pm^w$, where $n_\pm^w$ is the initial concentration of ionic species in pure water. Thus, the formula for the static dielectric constant of aqueous electrolytes is [17]:

$$\Delta\varepsilon^e = \Delta\varepsilon_w\left(1 - \frac{\left(c_0 - \alpha\gamma c_0^e\right)^2\beta c_0^e}{2c_0^2\left(K_2 + K'_2\beta c_0^e\right)}\right)^{1/3}, \tag{5.16}$$

where the coefficients have exactly the same meaning as in (5.15).

Figure 5.5 shows the dependencies of σ_{dc}^e and $\epsilon(0) = \Delta\epsilon^e + \epsilon_{THz}$ on the concentration c_0^e of the electrolyte according to (5.15) and (5.16). As one can see, the dependencies are similar to those observed in the experiment (see Figs. 5.2 and 5.4a). The best-fit parameters for HCl, NaOH, and NaCl to the experimental data are given in Table 5.1. In addition, the following parameters were used: c_0 = 55.5 mol/l as the initial concentration of H_2O molecules, and $\epsilon_w(0) = 78.5$ as the dielectric constant of pure water.

The model is generally valid for any strong electrolytes of the form MX. To apply the theory to other electrolyte systems, one could simply replace Na or Cl in

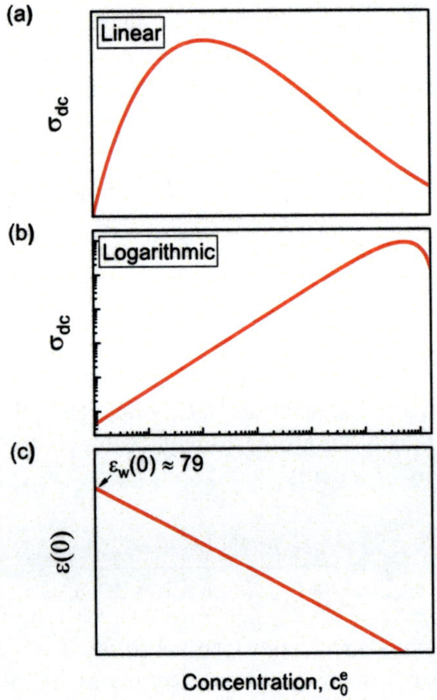

Fig. 5.5 The concentration dependencies of: **a**, **b** the DC conductivity σ_{dc}^{e} (logarithmic and linear scales), and **c** the static dielectric constant $\epsilon(0)$ of the aqueous HCl, NaOH, and NaCl solutions according to (5.15) and (5.16) (compare with Figs. 5.2 and 5.4a)

Table 5.1 The parameters of the best fit of (5.15) and (5.16) to the experimental data for aqueous solutions of HCl, NaOH, and NaCl

Parameter	HCl	NaOH	NaCl
Λ_0 ($\mathrm{S{\cdot}cm^2mol^{-1}}$)	426	248	126
D ($\mathrm{nm^2/ps}$)		9.3	
K_1		6.7×10^{-4}	
K_2 (mol/l)		1.2	
K_2'		0.1	0.2
K_4		–	0.1
α		2.0	1.4
β		1.0	0.6
γ	1.3	1.0	1.4

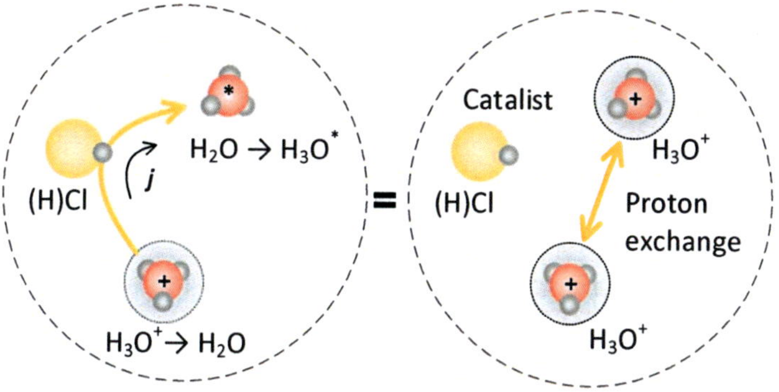

Fig. 5.6 The mechanism of the catalysis of the protonic conductivity in water by the species of electrolyte, which on large timescales (>1 μs) look like intermolecular proton exchange between intrinsic species of water, contributing to the DC conductivity of electrolyte solutions with lower activation energy than intrinsic ionic species in pure water

the formulas (5.4) to (5.16) with M or X, respectively. Note that the model above is a very simplified, as we made some assumptions regarding the initial species and the chemical products, and also about the symmetry of reactions (5.15) and (5.16). However, the curious could train themselves by implementing all the details of the model. Although this will significantly complicate (5.15) and (5.16) for the DC conductivity and the static dielectric constant, these formulas will describe the experimental data with better accuracy.

As one can see from Fig. 5.5a, the model satisfactorily reproduces the concentration dependence of the DC conductivity of electrolytes up to very high concentrations, including those that the standard Arrhenius model fails to reproduce. Note also that although models (5.15) and (5.16) are quite rough, they are significantly simpler than those provided by the Debye–Hückel corrections to the theory of electrolytic dissociation. Thus, the ionic model provides an alternative description of the electrodynamic properties of aqueous electrolytes without a complicated analysis of the hydration shells of ions. Instead, the ionic model accounts for local chemical reactions, which appear between the intrinsic ionic species of water and the species of the solute, thus, it is in agreement with the hypothesis of Mendeleev, who pointed out the importance of the chemical interaction between solute and solvent [8].

The central idea of the mechanism discussed above is that the intrinsic ions of water are too short-lived to contribute to the DC conductivity of water (see Chap. 4), however, when an electrolyte is added, the local chemical reactions change the interactions and the lifetime distribution of water species, catalyzing (effectively) the mobility of excess protons (see Fig. 5.6), and making them "visible" in DC conductivity. Thus, in contrast to the Arrhenius model, water is the source of charge carriers for DC conductivity, while the dissolved electrolyte is only a catalyst. Hence, the model takes into account the dynamic processes (local chemical reactions) on nanosecond and picosecond timescales and goes beyond the existing static models.

5.2 The Electrodynamics of Confined Water

In many natural and artificial systems, such as rocks, polymers, nanostructures, and biological objects, water is confined in small cavities bounded by a substance (liquid or solid) and exhibits properties different than those of bulk water. Such water can be considered as one more state of water: confined water. Confined water can be defined as water that is subject to geometric constraints on a nanoscale (typically a few nanometers) so that molecules are close enough to an interface to sense some difference from standard bulk conditions. Confined water is most frequently found in porous media or in solvation shells of colloidal particles. Since confined water is of high practical importance, it has been studied by means of different methods and techniques, and many parameters have been found to differ from those in the bulk water. In particular, low viscosity, a low dielectric constant, enhanced molecular mobility, surface charge, and shifted freezing and evaporation points have been observed experimentally or studied theoretically [31–37].

The electrodynamics of confined water is a matter of separate discussion, as it reflects the molecular mechanisms near the interfaces, and is also important for modern electric power storage systems, such as accumulators, flow batteries, supercapacitors, and fuel cells (see Sect. 5.6). For these systems, the properties of water near the interface are crucial and either serve as the reference parameters for the quantitative models or are used directly for the accumulation and separation of electric charges. Typically nuclei are separated from electrons (for example near the electrodes), but ion–ion separation also takes place (for example, in membranes and separators). However, the electrodynamics of water at nanoscales and near the interfaces is still poorly understood, in spite of the long history of its study, starting from the famous analysis of the electric double layer[8] by Helmholtz in 1853 [38].

The molecular mechanisms of the electrical charge separation near the aqueous interfaces (or in confined water) are important for processes on different scales, from the global [39] (see Sect. 5.3), to the microscopic in which the electric properties of confined water define the communication of cells, by providing neuronal-network pulses a propagation environment. Although the mechanism of the nerve pulse transfer has been considered from different angles, and the basic models are well established [40], the role of water in the charge-transfer mechanism in biological systems is underestimated. The picosecond timescale processes, discussed in Chap. 4, are very important for the electrodynamic properties of water near the interfaces and significantly influence local chemical reactions and transport in biological systems.

There are two basic parameters of interfacial water which characterize its electrodynamic properties: the static dielectric constant $\epsilon(0)$, and the protonic conductivity σ_{dc}. However, neither have been studied experimentally at length, because it is extremely difficult to separate the intrinsic properties of confined water from the properties of the confining matrix. Fumagalli et al. [32] studied the dielectric

---

[8] An electric double layer occurs when two phases, one of which is a liquid, come into contact. Trying to lower the surface energy, the particles at the interface form a different dynamical structure than that in the bulk.

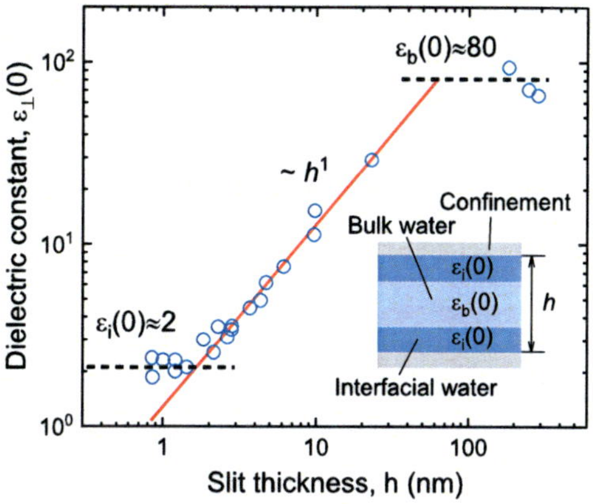

Fig. 5.7 The dielectric constant of water confined in a narrow slit channel as a function of the slit thickness h. The horizontal lines show the dielectric constants of bulk water (upper) and water at optical frequencies (lower). The inset shows the geometry of the sample, where two layers of interfacial water (dark blue) are separated by the layer of bulk water (light blue). The red is the linear fit. The experimental data points from [32]

constant of water under strong confinement and found that the dielectric constant of interfacial water is significantly lower than that for bulk water. They fabricated thin channels from 1 to 300 nm, which were filled with water. Measurements with an atomic force microscope showed that the dielectric constant of confined water is pore-size dependent as shown in Fig. 5.7. The bulk behavior ($\epsilon(0) \approx 80$) was observed only for the channels wider than 200 nm. In smaller pores, the dielectric constant changes approximately linearly with h and reaches a limiting value of about 2.1 at $h < 2$ nm. The model of multilayered water structure, as shown in the inset of the graph, allowed them to identify the presence of a near-surface layer with ϵ_i = 2.1 and thickness h_i = 7.4 Å, which was assigned to pure interfacial water. Note that the thickness h_i consists of only a few molecular layers of water and is close to the parameter L of the ionic model of water (see Table 4.2).

Fumagalli et al. suggested that the small ϵ_i is due to the inhibition of the rotational motion of water near the solid surface. The interaction of water molecules with the walls is assumed to reduce their polarizability. However, as discussed in Sect. 2.4.2 at temperatures far from absolute zero (the experiment was done at room temperature), the thermal energy is high enough to disorganize any molecular structures in liquids, even in strong confinement. Although the effect of the confining matrix on the structure of water is obvious, it is hard to believe that there is a permanent alignment of the molecular dipoles near the surface. Another explanation of the result is possible within the ionic model of water. Note that the parameter $L \approx 8$ Å of the ionic model, which is close to the observed thickness of the interfacial water layer, is associated with the average distance between short-lived spontaneously formed ionic species of water (see Sect. 4.2.3). In Sect. 4.5.3 we discussed that the static dielectric constant is formed by polarization due to relative displacement of excess protons and proton holes (short-lived H_3O^+ and OH^- ions) [41], rather than the rotation of molecular dipoles, which provide less than a 5% contribution to the static dielectric constant. Thus, the suppression of the dielectric constant of water in the pores, whose dimensions are less than L, can be associated with the spatial heterogeneity of water. The dielectric properties of water on a scale of less than the radius L of the ionic atmosphere of the intrinsic short-lived ions are different in comparison with the large-scale properties. In other words, the walls of the confining matrix prevent the formation of the atmosphere of counterions around the central ion, thus, eliminating the polarization unit, which is responsible for the polarization in the bulk. Note, however, that experiments with water in strong confinement show a different value of the dielectric constant than for bulk water against the wall of a container. In the latter case, as, for example, in the Leiden jar (see Sect. 5.6), the charges (H_3O^+ and OH^-) form the electric double layer, in which polarizability is very high. The dielectric constant of a double layer of water is many orders of magnitude larger than that for the bulk water.

Another important parameter of interfacial water is electric conductivity, which was studied in [42]. The model system was made of granular particles of nanodiamond of different diameters in such a way that it was possible to change the pore sizes from 1 to 100 nm. Figure 5.8 shows the dependence of the electrical conductivity of confined water on the pore size. The electrical conductivity increases in inverse proportion to the pore size and reaches maximum in the pores of about 2 nm, i.e., the same pore radius as was found for the minimal dielectric constant. The conductivity σ_i of the interfacial water layer was 0.02 S/cm, five orders of magnitude larger than that of bulk water. The optimal water layer for the maximal protonic conductivity was $h \approx 1 \pm 0.25$ nm, which is in agreement with the thickness h_i found above. Note that for the pore size less than h_i the conductivity goes down (see Fig. 5.8), which is presumably connected with the loss of percolation among water volumes between the grains.

The two experiments described above reveal the presence of an interfacial layer near aqueous interfaces with a vanishingly small polarization, but very high protonic conductivity. This layer is found to be a few molecules thick and has an electrical conductivity five orders of magnitude large than that of bulk water. Note also that σ_i exceeds the ionic conductivity of different superionic conductors in the temperatures range 0–100 °C (see Fig. 5.21). These results require interpretation, which satisfy both the low polarizability and the high protonic conductivity of the interfacial water layer.

Figure 5.9 shows the difference between bulk and interfacial water from the viewpoint of the ionic model of water (see Chap. 4). In bulk water, the mutual screening of short-lived spontaneously formed ionic species causes its high dielectric constant $\epsilon \approx 80$ (see Sect. 4.5.3) and low DC conductivity $\sigma_{dc} = 5.5\times10^{-8}$ S/cm. The first parameter results from the dipole moment μ (red arrows), induced by the relative displacement of ions, and the second is a result of the counteraction of the driving force, F_1, and the friction force, F_2. The latter is due to the electrostatic attraction of the ion by the center of its ionic atmosphere (the blue "cloud"). Interfacial water is characterized by the reduced ionic atmosphere, which is incomplete on one side,

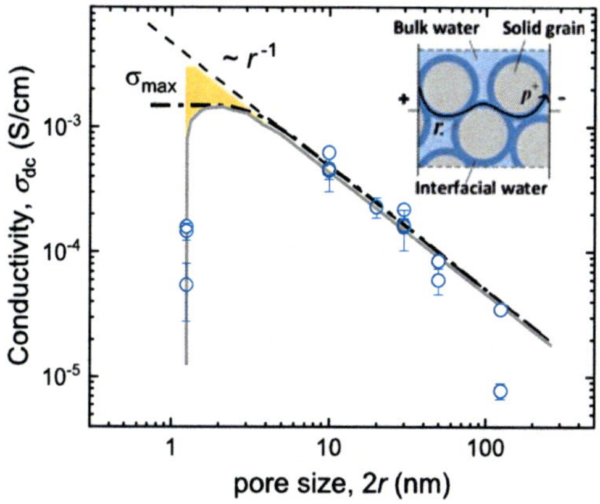

Fig. 5.8 The electric conductivity of water confined in a porous granular medium as a function of the pore radius r. The inset shows the structure of the sample: bulk water (light blue) coexists with the interfacial water layer (dark blue) around solid granular particles (gray). Black arrow shows the trajectory of the proton diffusion. The maximum conductivity value is observed in the pores of about 2 nm. Adapted with permission from [42]. Copyright 2020 American Chemical Society

Fig. 5.9 I-structures of **a** bulk water (ϵ = 80, low σ_{dc}), and **b** interfacial water ($\epsilon \approx 2$, high σ_{dc}). Colored circles are short-lived spontaneously formed ionic species of H_3O^+ (excess proton) and OH^- (proton hole). White open circles are neutral water molecules. The blue "cloud" is the ionic atmosphere. F_1 and F_2 are the driving and restoring (friction) forces, respectively. The red arrows are the effective dipole moments μ due to relative displacement of ions

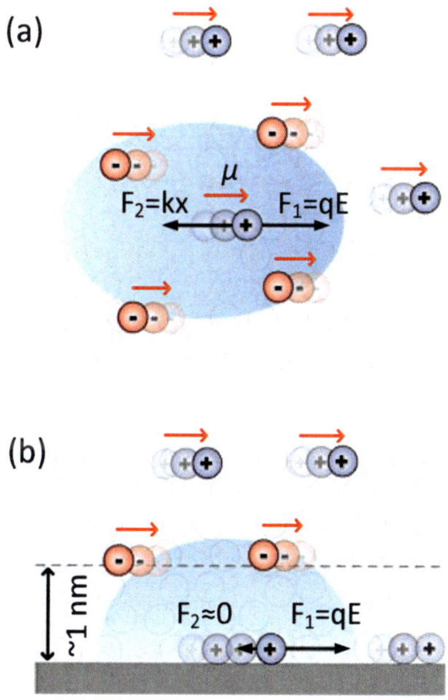

because of the wall of the confining material. As the ionic atmosphere is broken, there is no corresponding polarization, and, thus, results in the low dielectric constant $\epsilon_i \approx 2$. However, the ionic atmosphere, destroyed by the boundary, prevents the formation of an electrostatic restoring force, thus, the interfacial water layer possesses enhanced protonic conductivity. This interpretation of the enhanced protonic conductivity of the interfacial water layer and the suppressed dielectric is in a very good agreement with the experimental results discussed above.

In conclusion, there are many other interesting properties of interfacial water, which, however, go beyond the current book. From the viewpoint of electrodynamics, water in nanoconfinement exhibits properties of two types: that of bulk water and interfacial water, which differ in their protonic conductivity. Near the interface, the protons have higher mobility and are responsible for the enhanced conductivity in an external electric field. The intrinsic protonic conductivity of interfacial water, σ_i = 0.02 S/cm, is five orders of magnitude larger than what is observed for bulk water. This result provides a new insight into the fundamental properties of water and helps to better understand the general electrodynamic effects in porous membranes and other water-based systems in geological, biological, and engineered environments.

5.3 Atmospheric Electrodynamics and Aqueous Interfaces

The dielectric properties of water play an important role in atmospheric phenomena. The microscopic charge separation mechanism in clouds is laterally responsible for the weather on Earth. About 2,000 thunderstorms are active around the Earth right now. They cover an area of about 5×10^5 km^2, which is equal to the area of Spain. Figure 5.10 shows the global electric circuit. Thunderstorms behave like a voltage generator (or a battery) which is connected with the highly conducting ionosphere and Earth by the barely conducting lower and middle atmosphere. The potential difference between the Earth and the ionosphere is very high, and the strength of the electric field depends on the altitude. The typical electric field strength near the surface (in the environment where people spend most of their time) is about 120 V/m. The electric current flows back to Earth outside the thunderstorm region (see green arrows). Although the global circuit involves many interesting objects to study, the most exciting events appear in thunderstorms, which serve as an area of charge separation, lightning, and precipitation. All these effects are determined by the electrodynamic properties of water.

Let us consider the distribution of water in the atmosphere of Earth. The majority of atmospheric water is located in its lowest layer, the troposphere, where water exists as a gas, a liquid, or a solid. Atmospheric water vapor consists of molecular species but also contains small charged clusters $H_3O^+\cdot(H_2O)_n$ and $OH^-\cdot(H_2O)_n$, where $n \approx$ 4–12 with typical radii about 1 nm (nanodroplets), which exist even below saturation point. Interestingly, electrical conductivity measurements [44] show that moist air contains up to 10^{15} ions per liter of air, or about 10 parts per billion (ppb), which is high enough to affect the electrical properties of the atmosphere. If moist air becomes

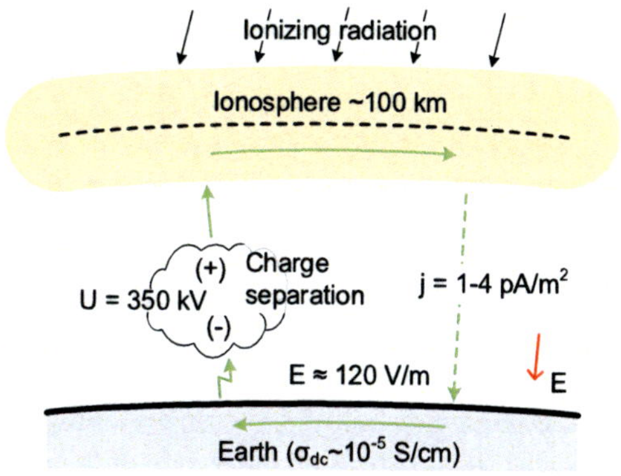

Fig. 5.10 The global electric circuit between the Earth and the ionosphere with a thunderstorm as a generator. Earth's atmosphere can be considered as a spherical capacitor, in which the first capacitor "plate" is the Earth surface, and the second "plate" is ionosphere. The average distance between the "plates" is about 100 km, and the voltage is about 350 kV

saturated, water molecules condense to droplets with a typical diameter from 10 to 100 μm, and larger rain droplets with typical radii of 0.1–1 mm. The latter are large enough to fall down to the Earth's surface, while the former cannot overcome the updraft flow of 5–10 m/s and go upward. Since the temperature at the top of the clouds reaches −50 °C, many water particles exist in the form of ice crystals, which traveling down to the lower warmer regions of the cloud may form aggregates (snowflakes) up to several centimeters long. Note that the phase transitions from vapor to liquid, and from liquid to solid release latent heat, which locally increases the temperature of the air. As a result, water in clouds remains liquid down to −20 °C.

Observations show that the electric charge in the clouds is carried mainly by the droplets and the ice crystals, the small cloud droplets are predominantly positively charged, the large rain droplets are mainly negatively charged. The amount of charge per droplet q depends on their size. Droplets with a diameter of 1 mm carry up to $q/e \approx 2\times10^7$ elementary charges [39]. Note that water is dielectric (see Chap. 1) and thus can polarize in an external electric field E in such a way that the secondary electric field E' of the polarized particles itself weakens the primary field of Earth. However, the unique properties of water allow the droplet to carry the charge, as a result of the separation of hydronium and hydroxyle ions. The upper limit of the charge a droplet can carry is reached when the secondary field E' equals the primary field E by magnitude (see Fig. 5.10). Thus, the maximal charge of the droplet is usually limited to about 10^{-11}C. [9]

The electric conductivity in the clouds is controlled by the local balance of sources and sinks [39]. The dominant sink is caused by cloud and aerosol particles, which reduce the electric conductivity within the clouds by more than a factor of 10 compared with the fair weather value. Conductivity decreases with an increasing electric field. However, when the field strength in the cloud is very large ($|E| > 300$ kV/m), corona discharge from water droplets or ice pellets, which ionize the air, may greatly enhance the ion production rate and thus locally compensating for the decrease of

[9] The net charge of Earth is about $|10^5|$ C, and about $|10^0|$ C including the charge of its atmosphere [43].

the conductivity [45]. The difference in conductivity between clear air and a cloud causes a layer of space charge to form on the boundary between the cloud and clear air if electric currents flow through that boundary. This surface charge produces a discontinuity in the electric field at the boundary. Thus, lightning discharges act like short circuits inside the clouds.

As mentioned above, the neutral water particle can polarize, but also needs a mechanism of charge separation to charge the global battery (see Fig. 5.10). There are several theories of the charge separation mechanism in the atmosphere [45], most of these theories are based on precipitation as the power mechanism of electrification. Figure 5.11 shows one of the charge separation mechanism in clouds. Initially, an external electric field E between the ionosphere and the Earth (see Fig. 5.10) induces the electric polarization of the rain droplet or the ice particle, which depends on its dielectric constant (see Sect. 2.4). Then, a smaller particle (another ice crystal or water droplet) gains the charge during a collision and exchanges protons or proton holes with the large particle. The typical time of the collision is about 10 ms. The particle which gains the positive charge is usually lighter and moves upwards within the updraft. The negatively charged rain droplets and ice particles descend due to their greater weight, thus enhancing the original electric field (Fig. 5.10). The amount of charge exchange between the large and small particles increases with an increasing electric field. The effect of the charge separation by collision is, thus, supported by a positive feedback loop that increases the original electric field until the limiting charge value of $q_{\max} = 4\pi\epsilon E r^2$, where r is the equivalent droplet radius, is reached.

The excess charge of the droplet or the ice particle is created by the intrinsic water ions (H_3O^+ and OH^-) (see Sect. 5.2). In the droplet, these ions are near the surface in conditions different from in the bulk (see Sect. 5.2). Positive and negative ions are screened in the volume of water, and the uncompensated charge is concentrated along the droplet surface being driven by Coulomb repulsion. The thickness h of the layer with excess charge can be determined by

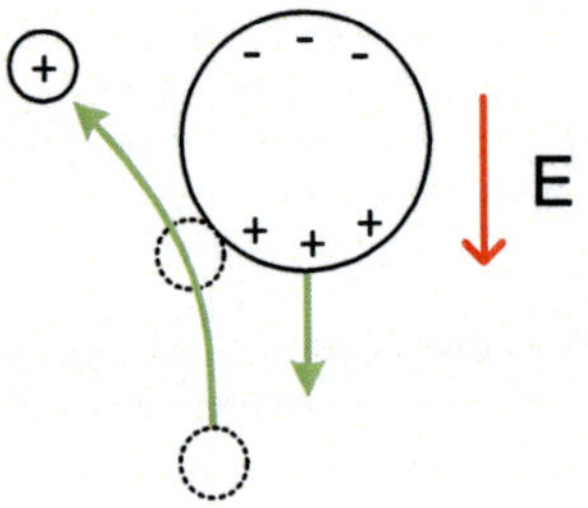

Fig. 5.11 The charge separation mechanism in a cloud. Collisions between two water/ice particles, polarized in the environmental electric field E, result in charge redistribution and separation. The smaller particle brings the positive charge up following updraft, and the larger particle brings negative charge down

$$h = \sqrt{\frac{\varepsilon_0 \varepsilon(0) \cdot k_B T \cdot \mu_\pm}{q \sigma_i}}, \tag{5.17}$$

where $\mu_\pm$ is the mobility of ions ($\mu_{H_3O^+} = 2.5 \times 10^{-7}$ m^2/V·s [12]), σ_i is the conductivity of interfacial water (see Sect. 5.2). Equation (5.17) gives $h \approx 1$ nm. This value does not depend on the droplet diameter; therefore, the specific charge of the droplet increases with a decrease in its diameter. In fact, most of the internal volume of the droplet remains neutral. However, the concentration of the charge near the droplet surface makes the charge/discharge of the droplet in collisions with other particles easier.

Despite their eminent importance, many processes discussed above, and their involvement in the continuous transformation between the liquid, solid, and vapor phases of water, the physical and chemical processes are not well understood on a fundamental level, thus hampering our predictive capabilities for climate change and water availability. Many of these unsolved puzzles are related to the interaction of water with aerosol particles, which offer an immense heterogeneous surface facilitating phase transformations and liquid- and gas-phase chemical reactions in the atmosphere. Particle porosity plays an important role in the interaction of water and atmospheric aerosols. For example, the specific area of illite mineral dust particles, a major component of desert dust [47], is 100 times larger than the surface area of a solid spherical particle of the same diameter. This property may preserve liquid water in the pores below the water saturation value, improving cloud condensation properties or influencing its efficiency in nucleating ice in supercooled water [48].

Recently, a mechanism of the heterogeneous deposition of ice nucleation came into the focus of the atmospheric research community. It was suggested that direct nucleation of ice from the vapor phase below water saturation is not likely to happen under realistic atmospheric cloud conditions and should be reconsidered in favor of a capillary condensation freezing mechanism, generally referred to as pore condensation and freezing [49]. Experimental and theoretical studies with model porous particles [50] have confirmed the feasibility of the mechanism, showing that homogeneous freezing in pores could account for some cases previously attributed to the deposition freezing mechanism.

Thus, the physicochemical properties of water and especially interfacial water are key to the most challenging questions in atmospheric science and geochemistry. Despite recent progress, our knowledge of the electric and structural properties of water near natural interfaces are still limited. Apart from the changed phase diagram of confined water, which is pore-size dependent, water and the walls of the porous material form complex interfaces characterized by an electrostatic potential, the ordering of molecular species, and a strong deviation of molecular kinetics from that of bulk water. Capillary condensation and retention, heterogeneous freezing, and heat transfer effects all depend on the properties of water near the interfaces, which are yet to be elucidated. Details of the dynamic structure of water, and its spatial-time heterogeneity, discussed in this book pave the way for a deeper understanding of the complex phenomena of atmospheric electricity and the corresponding natural phenomena.

5.4 Kelvin Water Dropper: Converting Gravity to Electricity

The interaction of droplets with electric fields was at the very beginning of the branch of science which we currently know as electrohydrodynamics. The pioneering works on this front were made by Rayleigh [51], and later by Taylor [52] and Melcher [53]. These works describe the basics of the complex phenomenon of the interaction of water with an external (static) electric field. In spite of the long history of study, the topic is still hot, and interesting engineering systems have continued to develop, utilizing the action of electric field on liquids in mass spectrometry, analytical chemistry, printing, coating, colloids propulsion, and even air purification.

An interesting effect of interaction of water with an electric field was demonstrated by Lord Kelvin, who described in 1867 a simple device that elegantly converts gravity to electricity without moving parts [54]. Figure 5.12 shows the schematic of the Kelvin dropper, which produces an electric potential difference by detaching the water droplets in the gravity field in a self-induced electric field.

The device consists of the two symmetric parts, the source of water has two nozzles, through which water drips under the force of gravity. If, owing to electrified bodies in the neighborhood, the potential in the air around the left-hand nozzle, where the stream breaks into droplets, is positive, the droplets fall with a negative charge, and vice versa for the right-hand nozzle. When the droplets fall into the receivers, they bring the electric charge, which is induced on the cross-wired metallic rings near the nozzles (the inducers). Suppose that a small negative charge is brought by the droplet from the left-hand nozzle to the first reserving container below. The inductor connected to this container becomes negatively charged, and the induced electric

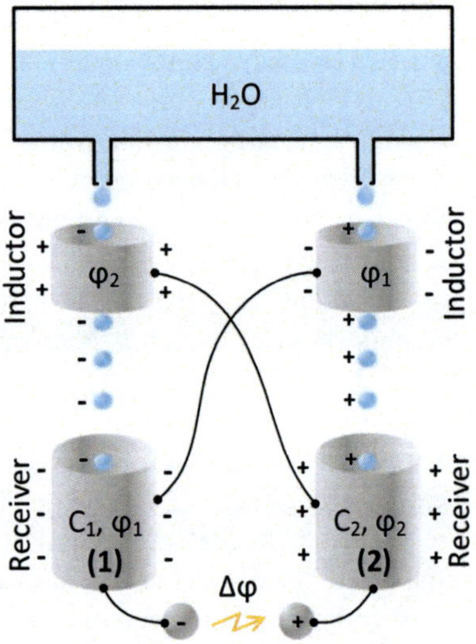

Fig. 5.12 The schematic of Kelvin electrostatic generator. Two falling streams of water droplets from the reservoir pass through metal rings (inductors), which are cross-wired, to receiving containers. See the text for explanation of the working principle

field near the second nozzle polarizes the droplet in such a way that its detached part is becoming positively charged. This positive charge, being brought to the second receiver, induces a positive charge in the ring, which stimulates negatively charged droplets from the first nozzle. Thus, subsequent cycles are enhanced by a positive feedback loop. As a result, the potential difference between receivers reaches several thousand volts, and one can observe periodic discharges, which bring the system back to the initial state.

Note that the induced voltage does not depend on the distance from the nozzle to the droplet receiver. In other words, the potential energy of the water tank, above the generator does not play a role, explaining the low efficiency of the Kelvin's generator. The only important part of the work performed by gravity is the detachment of the droplet from the rest of the water in the nozzle. As the potential of the receiver increases with time, the electric field of the inductor also produces the work which helps gravity to detach the droplets from the nozzle. Interestingly, measuring the charge of the droplets produced in the natural electric field of the atmosphere, Kelvin was able to measure the atmospheric potential with very high precision, comparable to modern plasma-based devices.

The working principle of the device can be described in the following way. Let C_1 and C_2 be the capacitances of the receivers (1) and (2), respectively (see Fig. 5.12), b_1 and b_2 are the rates of the charge loss per unit of time due to some leakage currents, and a_1 and a_2 are the rates of charge accumulation brought by the droplets. Let $+\varphi_1$ and $-\varphi_2$ be the time-dependent potentials of the receivers (1) and (2). Then, the action of the device is expressed by the following system of the coupled equations:

$$\begin{cases} C_1 \frac{d\varphi_1}{dt} = a_2\varphi_2 - b_1\varphi_1, \\ C_2 \frac{d\varphi_2}{dt} = a_1\varphi_1 - b_2\varphi_2. \end{cases} \tag{5.18}$$

Assuming that $C_{1,2}$, $a_{1,2}$, and $b_{1,2}$ are constants, and that $C_1 \approx C_2 = C$, $a_1 \approx a_2 = a$, and $b_1 \approx b_2 = b$, we get a simple solution for the potential difference between the receivers:

$$\Delta\varphi = \frac{a}{a+b} \exp\left(\frac{2(a-b)t}{C}\right) \tag{5.19}$$

which shows the exponential increase of the voltage with time. If $a = b$, the voltage does not change with time.

Figure 5.13 shows the droplet of water in an external electric field E_0. The electric field lines inside the droplet are roughly uniform, and the corresponding field $E' \approx E/\epsilon(0)$ is weakened by the value of the dielectric constant of water $\epsilon(0) \approx 80$. The field lines near the poles of the droplet are getting denser due to geometric factors. For a spherical droplet the pole field strength E_p is approximately three times larger than the strength of the external field E_0 and is determined by the formula:

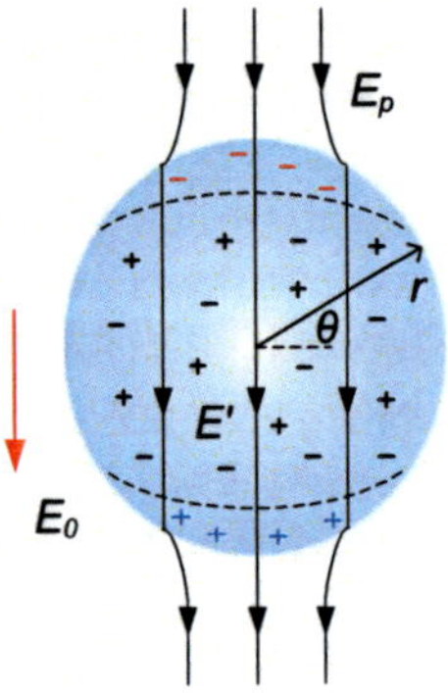

Fig. 5.13 The distribution of electric field lines around a charged droplet placed in a uniform electrostatic field E_0. E' is the strength of the resulting field inside the droplet. Pluses and minuses indicate the intrinsic ions of hydronium and hydroxyl ions, respectively. The poles, separated by the dashed lines, have excess charge

$$E_p = 3E_0 \cos\theta + \frac{q}{4\pi\varepsilon\varepsilon_0 r^2}, \tag{5.20}$$

where θ is the polar coordinate, and r is the radius of the droplet.

When an isolated droplet is subjected to a constant electric field E_0, the charges migrate to the poles of the droplet, being dictated by the direction of the field (see Fig. 5.13). The speed of the charge separation mechanism is dictated by the mobility of the intrinsic ions of water (see Table 1.1) and equals about 1 mm/s. The minimal electric field required for the complete separation of the H_3O^+ and OH^- ions in the droplet is $E_s = q/\epsilon\epsilon_0 S$, where q is the charge on the droplet poles, S is the area of the pole, $\epsilon(0)$ is the static dielectric constant of water, and ϵ_0 is the vacuum permittivity. For droplets about 0.1 mm in diameter, one has $E_s \approx$ 60,000 kV/cm, which is not reachable in the real experiment. Thus, the electric field does not separate charges but just polarizes the droplet. The relative displacement of the charges creates positively and negatively charged poles (see Fig. 5.13), while the droplet as a whole remains neutral. However, if the polarized droplet is separated into two parts along the external electric field, both parts carry the same electric charge, but with the opposite sign.

Figure 5.14 demonstrates the detachment of the water droplet from the nozzle in the external electric field. The nascent droplet has an elongated shape as shown in part (a) due to the simultaneous action of gravity and electrostriction. Then, before the droplet separates from the nozzle, a constriction is formed in its upper part as shown in part (b). After the detachment of the droplet, the constriction remains on the nozzle, and the detached part of the droplet acquires a shape close to spherical, as shown in part (c). The charge distribution in the direction of the electric field lines is not uniform. When the droplet forms, the intrinsic ions of water are displaced by the electric field in such a way that a part of the uncompensated charge appears near the droplet neck, and the oppositely charged part is formed in the lower part (tip) of the droplet. The droplet as a whole remains neutral until separation. After detachment of the droplet, an excess negative charge (see Fig. 5.14c) remains on the nozzle, while the excess opposite charge is carried away by the droplet. As a result, the spatial

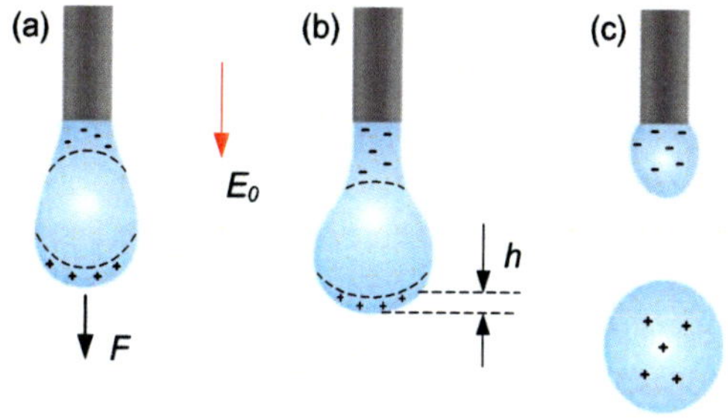

Fig. 5.14 The scheme of the droplet detaching from the nozzle in an external electric field E_0: **a** the droplet formation, **b** the droplet with a neck at the moment close to separation, and **c** the droplet separated from the nozzle. Pluses and minuses show uncompensated intrinsic ions of H_3O^+ and OH^-, respectively. The areas of uncompensated charge have the thickness of $h \ll 1\ \mu m$

separation of charges occurs under the action of gravity and the external field, which would not be achieved if the droplet polarization were caused by the orientation of H_2O dipoles.

To determine the dependence of the charge density $\rho_\pm$ (in C/m^2) on the electric field strength, we neglect the curvature of the excess charge area, whose thickness h defined by (5.17) is significantly smaller than the droplet diameter, and write

$$\rho_\pm = \nabla(\varepsilon\varepsilon_0 E_0) \approx \frac{\varepsilon_0 E_p}{h}, \tag{5.21}$$

where E_p is defined by (5.20). Equation (5.21) shows that the charge density is proportional to the applied field.

Thus, the electric field and gravity both separate the intrinsic H_3O^+ and OH^- ions of water. The separation of the hydronium and hydroxyl ions of water is the basic mechanism of the Kelvin electrostatic generator, in which the electrokinetic current depends on the overlap of the liquid velocity profile and the anisotropic charge distribution near the solid–liquid and liquid–air interfaces. As most of the water droplet does not carry the charge at the moment of its separation from the nozzle, smaller droplets could give higher efficiency. That is why, nanofluidic devices can be fabricated [55] based on a principle similar to that used in Kelvin's generator. The efficiency of the energy conversion of water-based systems, which is based on an acceleration/deceleration cycle, effected by jetting high-velocity charged droplets at a target with high electric potential [56], can reach up to 50%, which is higher than that for solar cell conversion systems (currently about 20%), and comparable with that for gas turbines. However, the standard liquid microjet device provides only about 10% efficiency of the energy conversion [57].

5.5 Water in a Strong Electric Field

An illustrative example of the anomalous behavior of water in an external electric field is the floating water bridge (Fig. 5.15): a stable filament between two beakers under high voltage. The phenomenon was demonstrated for the first time by Armstrong in 1893 [58] and has attracted attention in the past two decades [59–63] because the standard Bernal–Fowler model of water (see Chap. 1) fails to reproduce it. The effect consists in the increase of water's viscosity when an external electric field is applied, allowing one to create the conditions for the observation of the floating water bridge. Typically a voltage of about 15 kV is needed to form a stable free-hanging water wire of 1–2 cm. The effect has been shown [59] to be very sensitive to the purity of water: even small amount of foreign species destroy the bridge. The floating water bridge reveals the hidden properties of water, which are not observed in a zero electric field.

There are several competing phenomenological models that attempt to explain the bridge stability. These models are based on either notions of the dielectric constant ϵ or the surface tension γ [61–63]. Both are continual macroscopic parameters that lack microscopic clarity. As γ was shown [64] to be unable to hold the bridge alone, internal polarization-induced interactions were introduced [61] as forces which resist gravity. These forces originate from the intermolecular cooperativity effect, because the energy $A = E \cdot \mu = 10^{-7}$ eV of an interaction of dipoles of H_2O which are oriented along the field is five orders of magnitude lower than the thermal energy $k_B T = 10^{-2}$ eV; thus it cannot govern the bridge stability. Moreover, X-ray scattering and Raman spectroscopy showed [62] that there is no preferential orientation and no perturbation of water molecules along the field [65]. In other words, the alignment of bounded hyperpolarized H_2O molecules, which constitutes the Bernal–Fowler model of water (see Chap. 1), cannot explain the phenomenon of the bridge, but a comprehensive explanation can be found within the ionic model of water (see below).

While it is clear that the dielectric constant ϵ of water is an important parameter for the bridge stability, the microscopic origin of the excess polarization and, as a consequence, the cause of the internal force that acts against gravity must be elucidated.

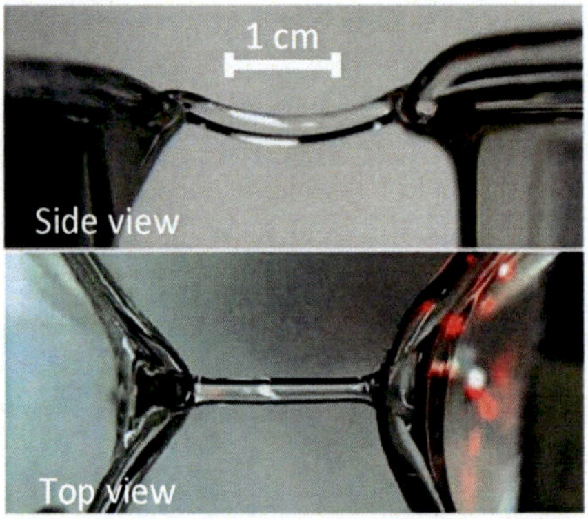

Fig. 5.15 The floating water bridge between two beakers with distilled water under voltage of about 15 kV. Reprinted with permission from [63] Copyright 2013 by the American Physical Society

Important results were recently obtained by ultrafast pump–probe spectroscopy [62] and neutron diffraction [66], which showed that the supramolecular structure inside the bridge is different than that observed in undisturbed water. Although there is no reason to expect anything else than just ordinary electrostatics, one should analyze the dynamic structure of water at high voltage in order to obtain more information on the microscopic level.

Fuchs et al. reported [67] a unidirectional transfer of both mass and charge through the bridge from anode to cathode. The intrinsic charges of water, being involved by the electric field, drag the neutral H_2O dipoles and carry the mass from the positively charged electrode to the negatively charged electrode. This direction corresponds to the direction of the excess-proton flow and can be understood in terms of electrophoresis.[10] Excess protons (H_3O^+) and proton holes (OH^-) as charges move in opposite directions, i.e., from anode to cathode, and from cathode to anode, respectively, but the mass (the hydrogen nuclei) always flows from anode to cathode, because both ions transfer the proton via Grotthuss mechanism (see Sect. 1.3.1). For instance, the dynamics of OH^- ions by the Grotthuss schematic assume that charge and the mass (of the proton) are moving in opposite directions, while in case of H_3O^+ the mass and the charge move in the same direction. Thus, the movement from anode to cathode of excess protons physically translate the mass from one beaker to another, while the charges move in the opposite directions. The unidirectional flow of protons through the bridge is confirmed [67] by the pH gradient between the beakers, which increases with time.

Namin et al. found [63] that the proton current is proportional to the applied voltage. But, when the voltage reaches 20–25 kV, the current saturates. This effect is similar to that observed in electrolytes, and known as Wien effect,[11] explained by Onsager and Kim [68]. In the ionic model of water (see Sect. 4.4), excess protons and proton holes have the potential to mutually interact. The external electric field tilts the potential, allowing them to drift along and against the field respectively, increasing the current density proportionally to the electric field. However, when the electric field strength becomes very high, the current saturates, as there is a limited amount of the ionic species in the potentials. The height of the potential determines the maximal current that can be passed through water without electrical breakdown.[12] Interestingly, the voltage of the saturation of the protonic current in the water bridge coincides with the threshold of electrostriction, observed in aqueous solutions [70], thus indirectly confirming that ionic species in water have the potential to mutually interact.[13]

---

[10] Electrophoresis is the motion of dispersed particles of a fluid under the influence of a spatially uniform electric field.

[11] The Wien effect is the experimentally observed increase in ionic mobility or conductivity of electrolytes at very high gradients of electrical potential [69].

[12] Electrical breakdown or dielectric breakdown is a process that occurs when an electrical insulating material (such as water), subjected to a high enough voltage, suddenly becomes an electrical conductor and a current flows through it.

[13] This interaction is missing in the standard Bernal–Fowler water model as discussed in Sect. 1.3.

Woisetschlager et al. found in addition [71] that the temperature of the bridge increases with time and finally causes its collapse after about 45 min by the stability breakdown. The effect can be interpreted by means of the electrophoretic retardation force, formed by the ions of the opposite signs, which moves in opposite directions under the conditions of electrostatic interaction. Note that the ionic model (see Sect. 4.2) assumes short-lived (picosecond) ionic species, with concentrations up to 1 mol/l, which is high enough to (a) explain the fast mass transfer through the water bridge observed in the experiment and (b) provide an additional electrostatic restoring force when the ionic species become relatively displaced by the external electric field (see below). The nanoscale density fluctuations inside the bridge (found by Skinner et al. [62]) indirectly confirm the 1 nm spatial heterogeneity, which was found in water by means of dielectric spectroscopy (see Chap. 4).

Figure 5.16 shows the scheme of the floating water bridge. The force T of internal tension counteracts to the gravitational force P. If the flexure h is much smaller than the distance between the beakers H, thus, the bridge lengths $l \approx H$. For the gravitational force, which acts on the each half of the bridge, one can write

$$P = SH\rho g/2, \tag{5.22}$$

where $S = \pi d^2/4$ is the cross-sectional area, d is the diameter of the bridge, ρ is the density of water, and g is the acceleration of gravity. Hereinafter we neglect the possible variation of the diameter d along the bridge. The tension force T can be defined as that with which the left and right parts of the bridge act on each other. At the suspension point B (see Fig. 5.16) the tension is equal to

$$T_B = \sqrt{T_x{}^2 + T_y{}^2} = T_O\sqrt{1 + 16h^2/H^2}, \tag{5.23}$$

where T_x and T_y are projections of T_B on the vertical and horizontal axes, respectively, and T_O is the tension at point O. For small h, $T_B \approx T_O \approx T$, which means that the tension T is approximately the same at any point along the bridge. The torques equilibrium dictates

$$T_O h = PH/4. \tag{5.24}$$

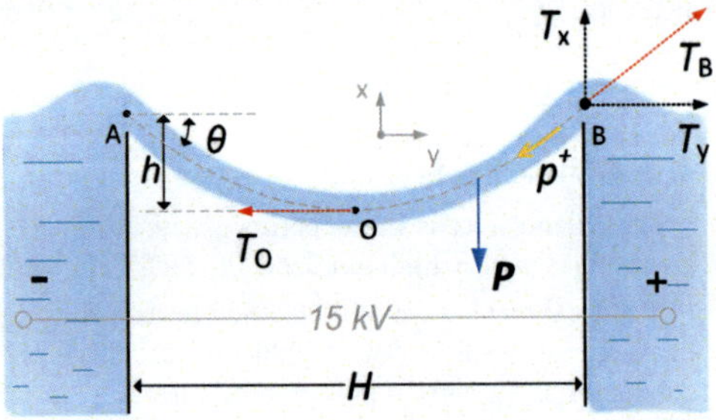

Fig. 5.16 The scheme of the floating water bridge between two beakers at high voltage. Arrows show the forces that act on the right-hand part of the symmetric bridge. The yellow arrow shows the direction of the proton flow as discussed in the text

Combining (5.22) and (5.24), we get the expression:

$$T = SH^2\rho g/8h, \tag{5.25}$$

which for $H = 2\,\text{cm}$, $h = 2\,\text{mm}$, and $d = 2\,\text{mm}$ (the real geometric parameters for the experimentally observed bridge) gives $T = 1.8\times10^{-3}$ N.

Figure 5.17 shows the distribution of excess protons and holes (H_3O^+ and OH^- ions) in the water inside the bridge. The external electric field tears the excess-proton dissociation–recombination trajectories and increases their lifetimes. As discussed in Sect. 4.2.3 (see Table 4.2), the maximal concentration of ionic species in water, both short- and long-lived is 1 mol/l (or 1 ion per about 50 water molecules). Perturbations caused by the high voltage should increase the concentration of long-lived ionic species $n_{\pm}$. These ionic species, according to the principle of the minimum entropy production, are distributed in such a way that each positively charged ion is surrounded by negatively charge ions and vice versa. Thus, each ion has its own ionic atmosphere. An external electric field E yields the shift Δx of charges from their equilibrium positions along and toward the field for H_3O^+ and OH^-, respectively (see Fig. 5.17). The restoring force F_c appears as a result of the attraction of the ions and the centers of their ionic atmospheres. This force in any cross section of the bridge equals

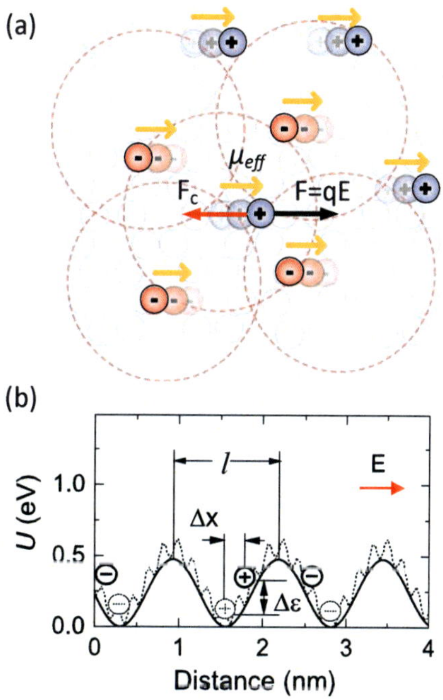

Fig. 5.17 The microscopic origin of the stability of the floating water bridge: **a** The structure of water inside the bridge. Color circles show the ionic species displaced by the external electric field. The red arrow shows the electrostatic restoring force. Yellow arrows show the effective dipole moments due to displacement. **b** The corresponding potential for ionic species

$$F_c = \kappa \Delta x \cdot d^2 (n_\pm)^{2/3}, \tag{5.26}$$

where $n_\pm$ is the concentration of ions, the term $d^2(n_\pm)^{2/3}$ represents the number of ionic species in the cross section of the bridge, and κ has the same origin as the constant $\kappa_1 = m_p \omega_0 \approx 0.04$ N/m, discussed in Sect. 3.5 in the context of the dielectric relaxation. The maximal possible relative displacement $\Delta x_{\max}$ of ionic species (see Fig. 5.17) equals the half the distance between them:

$$\Delta x_{\max} = l/2 = (n_\pm)^{-1/3}/2. \tag{5.27}$$

Substituting (5.27) for (5.26), we get $F_c = 60$ N for the bridge of d = 2 mm, where we assumed $n_\pm = 1$ mol/l from Table 3.5 as the maximum possible concentration of ionic species. One can see that the value of F_c significantly exceeds the tension force T found above. Thus, either not all ionic species are long-lived under the influence of the external electric field, or they are partially screened by the surroundings. Regardless, the internal restoring force, caused by intrinsic ionic species displaced by the field is high enough to explain the existence of the bridge. A more detailed analysis requires additional experimental data, which are currently missing. Nevertheless, we can make a few additional estimations, which confirm our suggestions.

The relative displacement of ionic species yields the polarization with the dipole moment $\mu_{\text{eff}} = q\Delta x$, where q is the effective charge of ions. All the induced dipoles are oriented in the same direction along the field (see Fig. 5.17), and, thus, attract each other with the force:

$$F_{dd} = \frac{q^2}{8\pi\varepsilon_0}\left[\frac{1}{l^2} - \frac{1}{(l-\Delta x)^2} - \frac{1}{(l+\Delta x)^2}\right] \cdot d^2 (n_\pm)^{2/3}. \tag{5.28}$$

Assuming again that $\Delta x = \Delta x_{\max}$, we get $F_{dd} = 300$ N, which is also more than enough to explain the stability of the bridge.

As, according to the ionic model, both the surface tension γ and the dielectric constant ϵ have the same microscopic origin: both result from the correlation of intrinsic ions of water, thus, both parameters are responsible for the bridge stability. This fact is in accordance with the experimentally observed coincidence in the temperature behavior between γ and ϵ [72].

Summarizing, the long-order electrostatic interaction of protons and holes forms a kind of plasma crystal (see Fig. 5.17), which does not form in undisturbed water. A quantitative estimate shows that the tear strength of the effective plasma crystal is high enough to resist the gravity. The phenomenon of the floating water bridge is a representative example of an effect which lies beyond the Bernal–Fowler model (see Chap. 1) but can be explained in the frame of the ionic model (see Chap. 4), showing the extended range of validity of the latter.

5.6 Water in Electrochemical Energy Systems

Electrochemical energy systems are the most rapidly growing area of energy sources, in which electrical energy is generated or accumulates by the charge separation in an entire material or at the interface by means of local chemical reactions, or by preliminary polarization [73–75]. Environmental friendly materials with high-energy density, high power density, and high electronic and ionic conductivity are in high demanded in this field. Thus, water and water-based systems attract a significant amount of interest as electrolyte materials. The study of these new materials, and the improvement of the technology of charge separation and conservation, constitutes the main challenge of modern electrochemistry [76].

Water has always played an important role in the electrochemical systems of the accumulation and generation of electricity. Such devices as batteries, fuel cells, supercapacitors, and electrolyzers, are either water based, or significantly depend on the electrodynamic properties of water. Water serves as an easily accessible, efficient, and ecologically friendly media for energy storage, because it is ubiquitous in nature, does not harm the environment, and the oxygen–hydrogen bond is one of the strongest chemical bonds (493 kJ/mol (5.1 eV)[14]). Even if water is not included in the electrochemical device directly, it is always present in or on the construction materials, changing their properties [77], and sometimes causing such unwanted problems as current leakage and self-discharge. In addition, water is a reference liquid for all aqueous electrolytes, whose dynamical structure determines the dynamic structure of any aqueous system. Thus, knowledge of the dielectric and structural properties of water in the bulk and the interfacial states are vital for the further development of electrochemical systems and beyond.

One of the first aqueous systems for the collection of electric charge was the Leiden (or Leyden) jar (see Fig. 5.18). This device, appearing in 1745, was connected to an electrophore machine and was able to store a high-voltage electric charge (up to 60,000 V, and $\sim$1 μC) between the inside and the outside surfaces of a glass jar filled with water. The Leiden jar was very popular for charge storage until the twentieth century, when electric circuits switched from DC to AC current and made the Leiden jar useless. Sometimes it is still used in electrotherapy and for educational purposes, continuing to amaze students with its unusual properties.

In order to find where exactly the charge is stored in a Leiden jar, Franklin created a modified version of it, with metal foil instead of a hand. He uses two metal plates and invented the first capacitor. He found [79] that both plates accumulate the charges of opposite signs of equal magnitudes. Then he removed the spike and poured out the "electrified" water. After that he filled the bottle with the new non-electrified water and found that jar was still storing the charge of almost the same magnitude as with the original electrified water. He concluded that charges are stored in the glass, not in the water.

[14]For comparison, the energy of O–O bond is 146 kJ/mol and H–H bond is 436 kJ/mol.

Fig. 5.18 The early Leiden jar (the first capacitor). The bottle with a metal spike through its stopper in contact with water. The hand serves as a first electrode, and the spike as a second, both separated by the dielectric (glass). The charges are stored in the film of water at the interface with glass. Open-access picture (originally from [78])

Later Addenbrooke showed [80] that Franklin's conclusion was wrong. He repeated the experiment with a "dissectible" jar made of three cylinders: two metal and one glass between them, which were in close contact. He charged the jar, dismounted the metallic electrodes, discharged them, and then assembled the jar again. The capacitor was charged as much as it had been originally. Thus, the charge seemed to be stored in the glass, as in Franklin's experiment. However, when he performed a similar experiment with a paraffin cylinder instead of a glass one, the result was the opposite: the reassembled capacitor was uncharged, and the charges were found on the metal cylinders before they were discharged. Addenbrooke concluded that the effect observed by Franklin is due to the water film, which always covers the glass surface. In other words, the charges are located in a film of water on glass, but not in the glass itself. When the conductor is removed (for example the water is drained), the charges remain in the water film on the glass surface. If the glass is carefully dried and the experiment is carried out in a dry atmosphere, then the "Franklin effect" is not observed. Thus, water in the Leiden jar is not as important as the interfacial layer (see Sect. 5.2), which is always present on the surface in standard conditions of temperature, pressure, and, most importantly, humidity. Disregarding the surface water sometimes leads to anomalous phenomena, which, however, have a simple explanation. Interestingly, the electrical double layer (about 1 nm thick) is determined by the intrinsic ionic species of H_3O^+ and OH^-, which determines the charge accumulation in such macroscopic systems as the Leiden jar and other similar devices.

While the Leiden jar was the first accumulator of electricity, the voltaic pile was its first source. Figure 5.19 shows the schematic of the pile assembled by Volta in 1794 that continuously provides an electric current to an electric circuit. Volta, relying on the earlier results by Galvani, found that when two metals are separated by cloth soaked with saltwater, they produce an electric current. Later Volta stacked several unit elements in parallel to increase the voltage. In this way he created the first battery. When the top and bottom contacts of Volta's battery are connected by a metal wire,

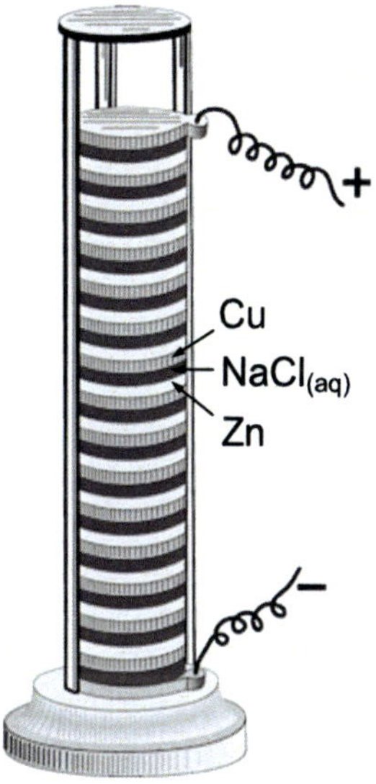

Fig. 5.19 A drawing of a copper–zinc voltaic pile similar to that assembled by Alessandro Volta in 1794. The copper and zinc discs are separated by spacers soaked in saltwater (the electrolyte), and assembled into the stack

an electric current flows through the pile and the connecting wire. The invention of the voltaic pile stimulated a series of scientific discoveries, such as the electrolysis of water,[15] the isolation of new chemical elements, electric discharges in gases and liquids, and oxidation–reduction or redox reactions.

Although Volta initially did not pay attention to the role of water in the contact with the metallic plates in his pile, later scientists realized that the aqueous electrolyte is involved in the chemical reactions with electrodes:

$$\begin{aligned} &\mathrm{Zn} + 2\mathrm{OH}^- \rightarrow \mathrm{Zn(OH)}_2 + 2e^-, \\ &2\mathrm{H}_3\mathrm{O}^+ + 2e^- \rightarrow \mathrm{H}_2 \uparrow + 2\mathrm{H}_2O, \end{aligned} \tag{5.29}$$

where the first reaction is oxidation, and the second is reduction. Thus, the battery provides an electrical current through an external circuit by means of chemical reactions: the zinc anode is oxidized by OH^- ions of water, releasing two negatively charged electrons. On the other side, two positively charged excess protons (H_3O^+ ions) of water accept two electrons from the metal, become reduced and form a neutral hydrogen molecule, which then volatilizes into the air. Note, that water plays the central role in the energy production, while NaCl (or any other electrolyte) just increases the rate of protonic transfer, supplying electrodes with the reaction components, e.g., ions of H_3O^+ and OH^- (see Sect. 5.1 for the dynamical structure of electrolytes in the frame of ionic model).

[15]The electrolysis of water is the decomposition of water into oxygen and hydrogen gases by means of electricity.

In the general case of a cell consisting of two metals A and B immersed in the solution of electrolytes AX and BX, the resulting electromotive force $\mathcal{E}$ is [81]:

$$e\mathcal{E}_{AB} = (\zeta_B - \zeta_{BX}) - (\zeta_A - \zeta_{AX}), \tag{5.30}$$

where ζ_A, ζ_B, ζ_{AX}, and ζ_{BX} are the chemical potentials of the metals A and B, and the corresponding electrolytes, respectively.

Although the Leiden jar and the Voltaic pile are not in high demand in modern engineering systems, they played an important role in the development of modern electrochemical energy systems. Being the simplest accumulator of electricity and generator, they served as prototypes of many modern devices, such as electrolytic capacitors, alkaline and acid batteries, flow batteries, supercapacitors, electrolyzers, hydrogen fuel cells, and metal hydride hydrogen storage materials, which are widely used for energy storage and conversion [29].

Figure 5.20 shows the general schematic for all modern electrochemical devices, which are used for the generation and accumulation of electricity. These devices have either open or closed systems, depending on the assembly and the purpose, with nearly the same operative principles; the only differences are in the structure, type, shape, and composition of the electrodes, electrolyte, and separator. The main difference between open and closed electrochemical systems is that the former are designed to generate electricity and require the continuous supply of reactants (fuel), while later just store energy in chemical reactions and are limited by the physicochemical properties of construction materials (electrodes and electrolyte). There

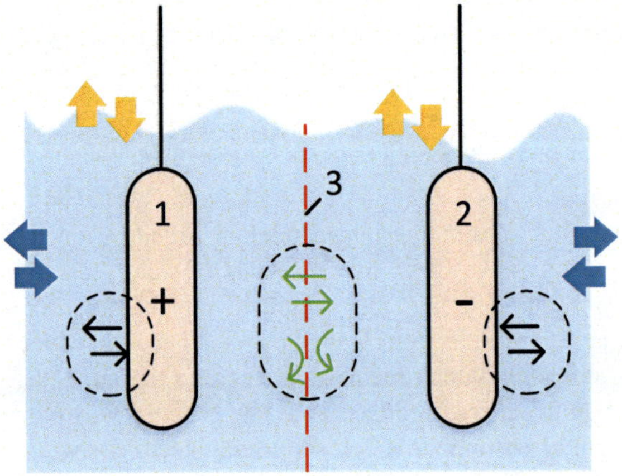

Fig. 5.20 The principle scheme of an electrochemical energy system (fuel cell, supercapacitor, battery, or electrolyzer). 1 and 2 are electrodes (cathode and anode), and 3 is the semi-permeable separator, immersed in the electrolyte. The cell can be closed (an accumulator), or open (a generator). Blue arrows show the inlet/outlet of the reaction components; yellow arrows show the exchange of the gas phase, or the charge leakage; black arrows indicate chemical reactions near the electrodes; and green arrows show behavior of ionic species through/near the separating membrane

are four main classes of the electrochemical systems: batteries, supercapacitors, fuel cells, and electrolyzers, which differ by their characteristics and thus ranges of application.

All batteries are variations of the galvanic cell described above. The most popular water-based battery is a lead–acid accumulator, which consists of a liquid electrolyte (a mixture of sulfuric acid and water), lead electrodes, and a separator (rubber or other porous material). The plates in the lead–acid battery contain an active material that is continuously bathed in the electrolyte while oxygen and hydrogen gas are released during charging. Due to the liquid convection and other effects described in this book, the water-based battery cell tends to rapidly charge and discharge. Having low power density (W/kg), but relatively high-energy density (W·h/kg), batteries are used to start automobiles, and in large backup power supplies, grid energy storage, off-grid household electric power systems, emergency lighting, and sometimes in small vehicles.

A special case of batteries is flow batteries, which are close to fuel cells (see below) by their operation principle. This type of electrochemical energy source stores electrical charge in scalable tanks of liquid electrolyte that is pumped near the electrodes to extract electrons by means of a chemical reaction. The used electrolyte returns to the tank and can be "recharged" by an external source of electricity. The two most popular, the organic and the vanadium flow batteries, are water based. Flow batteries are in demand as backup systems for stationary applications, such as wind power stations and solar-power plants. They have very low-energy density and relatively low charge and discharge rates, however, they can be easily scaled-up without a reduction of their efficiency.

Supercapacitors cover the gap between electrolytic capacitors and batteries. They store 100 times more energy per unit mass than electrolytic capacitors, accept and deliver charge much faster than batteries, and have very long charge-discharge life-cycles. The construction feature of a supercapacitor is electrodes of high porosity, and, thus, high surface area, so the electrolyte layer can be reduced to the thickness of only one electric double layer, which provides very high capacitance. The aqueous supercapacitor is very ecologically friendly as it contains only water and charcoal electrodes. Nevertheless, the threshold of the electrolysis of water limits the maximal voltage of such a supercapacitor to about 2 V. The properties of water presented in this book pave the way for the further improvement of aqueous supercapacitors. The main advantage of supercapacitors is that they have the highest power density among electrochemical energy sources, which makes them indispensable in automobiles and small planes, where they can effectively supplement fuel cells.

Fuel cells are open electrochemical systems which produce electric energy on demand using different types of fuels (e.g., hydrogen and natural gas). The most developed fuel cell type is an aqueous polymer-electrolyte fuel cell (PEMC).[16] This electrochemical system transforms the chemical energy liberated during the electro-

[16]Note that the solid-oxide fuel cell (SOFC) is one of the most promising types of the fuel cell for portable applications, as it has the highest energy density. However, the high operation temperature and the current level of material-production technology hinders its immediate application. The main

chemical reaction of hydrogen and oxygen to electrical energy, and consists of the electrodes, a proton or hydroxyl-conducting electrolyte, catalyst, and gas diffusion layers. The fuel cell requires a source of hydrogen gas, uses oxygen from the air, and produces water vapor as the product of the electrochemical reaction. Note that fuel cells are very efficient, but have the lowest power density among all electrochemical energy systems, thus, they should be used with supercapacitors, which together provide all the required parameters needed, for example, for electric vehicles. Thus, fuel cells are perspective competitors of the Li-ion batteries in the nearest future. In addition, a PEMC supplying an average household with electricity would also provide them with drinking water [82], which makes this type of device very ecologically friendly and suitable for indoor applications.

Finally, hydrogen, which is needed for the hydrogen fuel cell, is an intermediate product that stores electric energy to be converted into electric power on demand. The easiest and most efficient way to produce the hydrogen of very high purity (up to 99.999%) is the electrolysis of water. The composition of the modern electrolyzer is similar to that shown in Fig. 5.20, with some additional features, such as stirrers and ultrasonic generators, which increase the efficiency. Alkaline electrolyzers operate via the transport of hydroxide ions (OH^-) through the electrolyte from the cathode to the anode with hydrogen being generated on the cathode side. Hydrogen gas released in this way can be used as hydrogen fuel, or remixed with the oxygen to create oxyhydrogen gas, which is used in welding and other applications. Sometimes called water splitting, electrolysis requires a minimum potential difference of 1.23 V. However, as discussed in Sect. 5.2, the properties of water can be significantly changed near the interfaces, which can reduce the threshold of the electrolysis. Moreover, recent studies of the nuclear quantum effects in protic systems and their role in polarization effects [83] show that the mechanism of autodissociation in water needs to be revised, taking into account short (sub-femtosecond) processes and the spatial heterogeneity of water (see Chaps. 3 and 4).

Note that for all the systems described above, the most important parameter of water is the electric conductivity, or ability to transfer the electric charge. Figure 5.21 shows the temperature dependencies of different ionic conductors. As one can see water and ice in their bulk states shows negligibly small electrical conductivity in comparison with the main superionic conductors. However, as demonstrated in [41], and discussed in Sect. 5.2, interfacial water has an electric conductivity five orders of magnitude higher than that in bulk water, whose absolute value exceeds the conductivity of other electrolytes in an extended frequency range around room temperature. The remarkable properties of water at the nanoscale pave the way for the development of proton exchange membranes and aqueous electrolytes with better characteristics than are currently used. The increase of the proton-transfer rate will proportionally increase the power density of the corresponding electrochemical devices, which should, in particular, increase the efficiency and reduce the cost of electricity.

---

players in this market still use PEMC, however, there are many reasons to expect that SOFC will oust other fuel cell types in the near future.

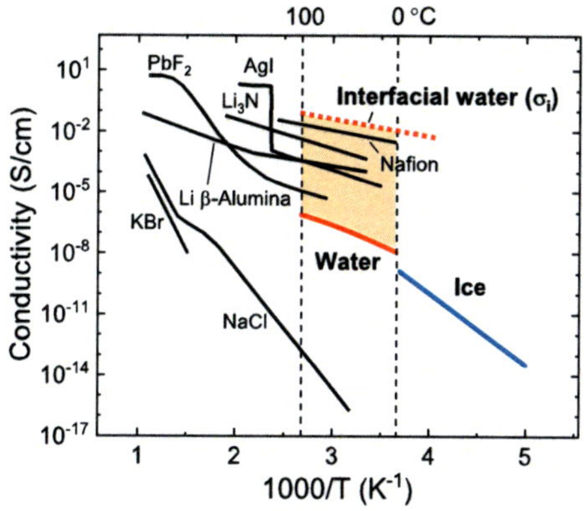

Fig. 5.21 The electric conductivity of ionic conductors as a function of inverse temperature. Red and blue lines denote the experimental curves for bulk water and ice, respectively. The dashed red curve is the upper limit of confined water conductivity and corresponds to pure interfacial water, expected for the materials with narrow pores (about 2 nm) and high porosity. The water-filled porous materials exhibit the values of conductivity inside the yellow-shaded region. Reprinted with permission from [42]. Copyright 2020 American Chemical Society

Further progress in the field of electrochemical energy generation and storage will significantly depend on our understanding of the dynamics and structure of water and aqueous soft-matter materials (such as porous membranes, water–solid and water–air interfaces, and colloidal suspensions) on the nanoscopic and macroscopic levels at short (picosecond) and relatively long (millisecond or longer) time intervals. The results discussed in this book serve as a foundation, which can be used to accelerate and simplify the further search for the appropriate materials for electrochemical needs, reveal the scope of the knowledge of the dynamic structure of water for a deeper understanding of environmental and biological systems. Apart the development of materials with high-energy accumulation density, I also expect that soon we will be able to create artificially charge-transfer systems based on protonic (not electronic) systems. This new branch of science, protonics, promises new bio-compatible charge and information transfer systems, which are free of the metal-dielectric interfaces and remove the problems of transfer between electronic and protonic systems.

References

1. J.W. Rodger, The electric conductivity of pure water. Nature **51**, 42–43 (1894)
2. W. Nernst, *Theoretische Chemie Vom Standpunkte Der Avogadro'schen Regel Und Der Thermodynamik: F* (Enke, 1900)
3. F. Kohlrausch, Einfache Methoden und Instrumente zur Widerstandsmessung insbesondere in Elektrolyten. Annalen der Physik. **11**, 653–660 (1880)
4. F. Kohlrausch, Uber die Wirkung der Polarisation auf alternierende Strome und uber einen Sinusinduktor. Annalen der Physik, 290–303 (1874)

5. S.A. Arrhenius, Recherches sur la conductibilité galvanique des électrolytes (Investigations on the galvanic conductivity of electrolytes), 1st edn. (Norstedt, Stockholm (1884)
6. D.I. Mendelejew, Das spezifische Gewicht der Schwefelsäure-Lösungen. Zeitschrift für Physikalische Chemie. **1**, 273–284 (1887)
7. D.I. Mendeleev, *The study of aqueous solutions by specific gravity* (St, Petersburg, 1887). [in Russian]
8. D.I. Mendeleev, Note on dissociation of solutes. J. Russ. Phys. Chem. Soc. **21**, 213–4 (1889). [in Russian]
9. S. Arrhenius, Hydration versus electrolytic dissociation. Phyl. Mag. **28**, 30–38 (1889)
10. P. Debye, E. Hückel, The theory of electrolytes. Physikalische Zeitschrift **24**, 185–206 (1923)
11. J.-J. Max, V. Gessinger, C. van Driessche, P. Larouche, C. Chapados, Infrared spectroscopy of aqueous ionic salt solutions at low concentrations. J. Chem. Phys. **126**, 184507–14 (2007)
12. P. Atkins, J. de Paula, *Physical Chemistry*, 9th edn. (W. H. Freeman and Company, New York, 2010)
13. A.A. Abrikosov, *Fundamentals of the theory of metals* (Elsevier Science Pub. Co, NY, 1988)
14. J. B. Hasted, D. M. Ritson, C. H. Collie, Dielectric properties of aqueous ionic solutions. Parts I and II. J. Chem. Phys. **16**, 1–21 (1948)
15. J.B. Hasted, *Aqueous Dielectrics* (Chapman and Hall, London, 1973)
16. C.P. Richter, A. MacLean, Salts taste threshold of humans. Am. J. Physiol. **126**, 1–6 (1939)
17. V.G. Artemov, A.A. Volkov, N.N. Sysoev, A.A. Volkov, Conductivity of aqueous HCl, NaOH and NaCl solutions: Is water just a substrate? EPL **109**, 26002–6 (2016)
18. R. Buchner, G.T. Hefter, P.M. May, Dielectric relaxation of aqueous NaCl solutions. J. Phys. Chem. A. **103**, 1–9 (1999)
19. G. Haggis, J. Hasted, T. Buchanan, The dielectric properties of water in solutions. J. Chem. Phys. **20**, 1452–1465 (1952)
20. E. Glueckauf, Bulk dielectric constant of aqueous electrolyte solutions. Trans. Faraday Soc. **60**, 1637–1645 (1964)
21. J. Liszi, A. Felinger, E. Kristof, Static relative permittivity of electrolyte solutions. Electrochim. Acta. **33**, 1191–1194 (1988)
22. H. Falkenhagen, *Electrolytes* (Oxford University Press, London, 1934)
23. V.V. Shcherbakov, Y.M. Artemkina, N.V. Pleshkova, K.R. Seddon, Ultimate high-frequency conductivity of solvent and electroconductivity of electrolyte solutions. Russ. J. Electrochem. **45**, 922–924 (2009)
24. J. Mähler, I. Persson, A study of the hydration of the alkali metal ions in aqueous solution. Inorg. Chem. **51**, 425–438 (2012)
25. M.N. Rodnikova, Negative hydration of ions. Russ. J. Electrochem. **2003**(39), 192–197 (2003)
26. J. O'M. Bockris, A. K. N. Reddy, M. Gamboa-Aldeco, *Modern Electrochemistry*, 2nd edn. (Springer, Berlin, 1998)
27. R.A. Robinson, R.H. Stokes, *Electrolyte Solutions*, 2nd edn. (Butterworths Scientific Publications, London, 1959)
28. S. Glasstone, *An Introduction to Electrochemistry*, 4th edn. (East West Press PVT, New Delhi, 1974)
29. V.N. Bagotsky, *Fundamentals of Electrochemistry*, 2nd edn. (Hoboken, New Jersey, John Wiley and Sons, 2006)
30. M.R. Wright, *An Introduction to Aqueous Electrolyte Solutions* (John Wiley and Sons, Chichester, 2007)
31. T.H. van der Loop, N. Ottosson, T. Vad, W.F.C. Sager, H.J. Bakker, S. Woutersen, Communication: slow proton-charge diffusion in nanoconfined water. J. Chem. Phys. **146**, 131101–5 (2017)
32. L. Fumagalli, A. Esfandiar, R. Fabregas, S. Hu, P. Ares, A. Janardanan, Q. Yang, B. Radha, T. Taniguchi, K. Watanabe, G. Gomila, K.S. Novoselov, A.K. Geim, Anomalously low dielectric constant of confined water. Science **360**, 1339–1342 (2018)
33. T. Seki, S. Sun, K. Zhong, C.-C. Yu, K. Machel, L.B. Dreier, E.H.G. Backus, M. Bonn, Y. Nagata, Unveiling heterogeneity of interfacial water through the water bending mode. J. Phys. Chem. Lett. **10**, 6936–6941 (2019)

34. H. Chinen, K. Mawatari, Y. Pihosh, K. Morikawa, Y. Kazoe, T. Tsukahara, T. Kitamori, Enhancement of proton mobility in extended-nanospace channels. Angew. Chem. Int. Ed. **51**, 3573–3577 (2012)
35. X. Ling, M. Bonn, K.F. Domke, S.H. Parekh, Correlated interfacial water transport and proton conductivity in perfluorosulfonic acid membranes. PNAS **116**, 8715–8720 (2019)
36. M. Prats, J. Teissié, J.-F. Tocanne, Lateral proton conduction at lipid-water interfaces and its implications for the chemiosmotic-coupling hypothesis. Nature **1986**(322), 756–758 (1986)
37. J. R. Macdonald, C. A. Barlow (Jr), Theory of double-layer differential capacitance in electrolytes. J. Chem. Phys. **36**, 3062–3080 (1962)
38. H. Helmholtz, Ueber einige Gesetze der Vertheilung elektrischer Ströme in körperlichen Leitern mit Anwendung auf die thierisch-elektrischen Versuche. Annalen der Physik und Chemie **165**, 211–233 (1853)
39. H. Volland, *Atmospheric Electrodynamics* (Springer, Berlin, 1984)
40. J.A. Anderson, E. Rosenfeld, *Neurocomputing: Foundations of Research* (The MIT Press, Cambridge, Massachusetts, 1988)
41. V.G. Artemov, E. Uykur, P.O. Kapralov, A. Kiselev, K. Stevenson, H. Ouerdane, M. Dressel, Anomalously high proton conduction of interfacial water. J. Phys. Chem. Lett. **11**, 3623–3628 (2020)
42. V.G. Artemov, E. Uykur, S. Roh, A.V. Pronin, H. Ouerdane, M. Dressel, Revealing excess protons in the infrared spectrum of liquid water. Sci. Rep. **10**, 11320–9 (2020)
43. M.A. Uman, The Earth and its atmosphere as a leaky spherical capacitor. Am. J. Phys. **42**, 1033–1035 (1974)
44. H.R. Carlon, Ion content of air humidified by boiling water. J. Appl. Phys. **51**, 171–173 (1980)
45. M.R. Pruppacher, J.D. Klett, *Microphysics of Clouds and Precipitation*, 2nd edn. (Springer, Berlin, 2010)
46. A.N. Aufdermauer, D.A. Johnson, Charge separation due to riming in an electric field. Q. J. R. Meteorol. Soc. **98**, 369–382 (1972)
47. I.N. Sokolik, O.B. Toon, Incorporation of mineralogical composition into models of the radiative properties of mineral aerosol from UV to IR wavelengths. J. Geophys. Res. **104**, 9423–9444 (1999)
48. R. Wagner, A. Kiselev, O. Moehler, H. Saathoff, I. Steinke, Pre-activation of ice-nucleating particles by the pore condensation and freezing mechanism. Atmos. Chem. Phys. **16**, 2025–2042 (2016)
49. A. Kiselev, F. Bachmann, P. Pedevilla, S.J. Cox, A. Michaelides, D. Gerthsen, T. Leisner, Active sites in heterogeneous ice nucleation–the example of K-rich feldspars. Science **355**, 367–371 (2017)
50. R.O. David, C. Marcolli, J. Fahrni, Y. Qiu, Y.A. Perez-Sirkin, V. Molinero, F. Mahrt, D. Brühwiler, U. Lohmann, Z.A. Kanji, Pore condensation and freezing is responsible for ice formation below water saturation for porous particles. PNAS **116**, 8184–8189 (2019)
51. L. Rayleigh, On the equilibrium of liquid conducting masses charged with electricity. Philos. Mag. **14**, 184–186 (1882)
52. G. I. Taylor, Disintegration of water drops in an electric field. Proc. R. Soc. London, Ser. A, 280, 383–397 (1964)
53. J.R. Melcher, G.I. Taylor, Electrohydrodynamics: a review of the role of interfacial shear stresses. Annu. Rev. Fluid Mech. **1**, 111–146 (1969)
54. W. Thomson, On a self-acting apparatus for multiplying and maintaining electric charges, with applications to illustrate the voltaic theory. Proc. R Soc. Lond. **16**, 391–396 (1867)
55. A.G. Marín, W. van Hoeve, P. García-Sánchez, L. Shui, Y. Xie, M.A. Fontelos, J.C.T. Eijkel, A. van den Berg, D. Lohse, The microfluidic Kelvin water dropper. Lab Chip. **13**, 4503–4506 (2013)
56. Y. Xie, D. Bos, L.J. de Vreede, H.L. de Boer, M. J. van der Meulen, M. Versluis, A.J. Sprenkels, A. van den Berg, J.C.T. Eijkel, High-efficiency ballistic electrostatic generator using microdroplets. Nat. Comm. **5**, 3575–5 (2014)

57. A.M. Duffin, R.J. Saykally, Electrokinetic power generation from liquid water microjets. J. Phys. Chem. C **112**, 17018–17022 (2008)
58. W. Armstrong, Electrical Phenomena. The electrical. Engineer **10**, 154–155 (1893)
59. E.C. Fuchs, J. Woisetschläger, K. Gatterer, E. Maier, R. Pecnik, G. Holler, H. Eisenkölbl, J. Phys. D Appl. Phys. **40**, 6112–4 (2007)
60. R.C. Ponterio, M. Pochylski, F. Aliotta, C. Vasi, M.E. Fontanella, F. Saija, Raman scattering measurements on a floating water bridge. J. Phys. D Appl. Phys. **43**, 175405–175408 (2010)
61. A. Aerov, Why the water bridge does not collapse. Phys. Rev. E. **84**, 036314–4 (2011)
62. L.B. Skinner, C.J. Benmore, B. Shyam, J.K.R. Weber, J.B. Parise, Structure of the floating water bridge and water in an electric field. PNAS **109**, 16463–16468 (2012)
63. R.M. Namin, S.A. Lindi, A. Amjadi, N. Jafari, P. Irajizad, Experimental investigation of the stability of the floating water bridge. Phys. Rev. E. **88**, 033019–6 (2013)
64. A. Widom, J. Swain, J. Silverberg, S. Sivasubramanian, Y.N. Srivastava, Theory of the Maxwell pressure tensor and the tension in a water bridge. Phys. Rev. E. **80**, 016301–7 (2009)
65. F. Saija, F. Aliotta, M.E. Fontanella, M. Pochylski, G. Salvato, C. Vasi, R.C. Ponterio, Communication: an extended model of liquid bridging. J. Chem. Phys. **133**, 081104–4 (2010)
66. E.C. Fuchs, B. Bitschnau, J. Woisetschläger, E. Maier, B. Beuneu, J. Teixeira, Neutron scattering of a floating heavy water bridge. J. Phys. D. Appl. Phys. **42**, 065502–4 (2009)
67. E. C. Fuchs, L. L. F. Agostinho, M. Eisenhut, and J. Woisetschläger, Mass and charge transfer within a floating water bridge. Proc. SPIE 7376, Laser Applications in Life Sciences, 73761E (2010)
68. L. Onsager, S.K. Kim, Wien effect in simple strong electrolytes. J. Phys. Chem. **61**, 198–215 (1957)
69. R. A. Robinson, R. H. Stokes, *Electrolyte Solutions. Butterworth and Co*, 2nd edn. (Courier Dover Publications 2002)
70. J.E. Desnoyers, R.E. Verrall, B.E. Conway, Electrostriction in aqueous solutions of electrolytes. J. Chem. Phys. **43**, 243–250 (1965)
71. J. Woisetschlager, K. Gatterer, E.C. Fuchs, Experiments in a floating water bridge. Exp. Fluids **48**, 121–131 (2010)
72. H.A. Papazian, Correlation of surface tension between various liquids. J. Am. Chem. Soc. **93**, 5634–5636 (1971)
73. J.B. Goodenough, K.-S. Park, The Li-Ion rechargeable battery: a perspective. J. Am. Chem. Soc. **135**, 1167–1176 (2013)
74. K. Chen, S. Song, F. Liua, D. Xue, Structural design of graphene for use in electrochemical energy storage devices. Chem. Soc. Rev. **44**, 6230–6257 (2015)
75. E. McCalla, A.M. Abakumov, M. Saubanére, D. Foix, E.J. Berg, G. Rousse, Visualization of O-O peroxo-like dimers in high-capacity layered oxides for Li-ion batteries. Science **350**, 1516 (2015)
76. K. Chen, D. Xue, Materials chemistry toward electrochemical energy storage. J. Mater. Chem. A. **4**, 7522 (2016)
77. M. A. Pogosova, I. V. Krasnikova, A. O. Sanin, S. A. Lipovskikh, A. A. Eliseev, A. V. Sergeev, K. J. Stevenson, Complex investigation of water impact on Li-Ion conductivity of $Li_{1.3}Al_{0.3}Ti_{1.7}(PO_4)_3$—Electrochemical, chemical, structural, and morphological aspects. Chem. Mater. **32**, 3723–3732 (2020)
78. W.J. Harrison, C.A. White, *Magnetism and Electricity*, 5th edn. (UK, Blackie and Son, 1898)
79. K. Kuehn, *Müschenbroek's Wonderful Bottle. A Student's Guide Through the Great Physics Texts*, pp. 43–60 (2015)
80. G. L. Addenbrooke, A study of Franklin's experiment on the Leyden jar with movable coatings. Philos Mag. 6th Series. 43, 489–493 (1922)
81. L.D. Landau, E.M. Lifshitz, *Electrodynamics of continuous media* (Pergamon Press, New York, 1958)
82. J.E. Tibaquirá, K.D. Hristovski, P. Westerhoff, J.D. Posner, Recovery and quality of water produced by commercial fuel cells. Int. J. Hydr. Ener. **36**, 4022–4028 (2011)
83. V.G. Artemov, A. Ryzhov, E. Carlsen, P. Kapralov, H. Ouerdane, Nonrotational mechanism of polarization in alcohols. J. Phys. Chem. B. **124**, 11022–11029 (2020)